Having Been A Soldier

THE AUTHOR
SEPTEMBER 1968

Having Been A Soldier

BY

LIEUTENANT-COLONEL

COLIN MITCHELL

*Every man thinks meanly of himself for not
having been a soldier, or not having been to sea.*
SAMUEL JOHNSON
10th April 1778

HAMISH HAMILTON
LONDON

First published in Great Britain, 1969
by Hamish Hamilton Ltd
90 Great Russell Street, London WC1
Copyright © 1969 by Colin Mitchell

SBN 241 01722 x

Printed in Great Britain by
Western Printing Services Ltd, Bristol

ILLUSTRATIONS

The Author   *frontispiece*

v

MAPS

FOREWORD

IT IS NOT unusual for retired military officers to write their memoirs. The bookshelves hold many such volumes, written by men past their prime but full of years and honour. They are the basis of much of our study of warfare and the personalities in command.

My own quarter of a century in the British Army ended prematurely and there is little precedent to justify a volume of military memoirs from one who did not attain the heights to which he aspired professionally. But as my resignation arose from the peculiar circumstances which surrounded the actions of the 1st Battalion Argyll and Sutherland Highlanders in Aden in 1967, and coincided with the decision of the Government to disband the Argylls by 1971, I believe that it is in the public interest to put the record straight.

This story of life in a famous regiment should highlight the need to retain the Argylls in the British Army—for the Argylls are the original *Thin Red Line* and it is inconceivable to think of the British Infantry without them; not so much because of the magnificent past of the Argylls but because of their high professionalism at the present time which attracts so many volunteers to their ranks when overall recruiting figures are low.

ACKNOWLEDGMENTS

I am immeasurably grateful to Tom Pocock who did so much to help with the preparation of the book and who did all the hard work serialising the draft while I went off to Vietnam as a war correspondent. I must also thank Richard Hough of Hamish Hamilton Ltd, who oiled the machine whenever it creaked and who did so much to promote the idea.

To reconstruct the events of 1967 in Aden I had the unstinted help of my comrades-in-arms in the 1st Argylls. They shared the heat and burden of the day in every sense. In particular I should mention Major Alastair Howman MBE, Major Ian Mackay MBE, MC, and Captain David Thomson MC.

My thanks are due to Terry Fincher and the *Daily Express* for allowing me to reproduce so many of his excellent photographs and to Boris Weltman for drawing the maps.

I have also to thank Lady Tweedsmuir for permission to quote from John Buchan's *Homilies and Recreations*, Macmillan & Co. for permission to quote from Eric Linklater's *Crisis in Heaven* and the Society of Authors as the literary representative of the Estate of A. E. Housman and Jonathan Cape Ltd., publishers of *The Collected Poems of A. E. Housman*, for permission to quote from 'Here dead lie we'.

CHAPTER 1

'Here dead lie we because we did not choose
To live and shame the land from which we sprung.
Life, to be sure is nothing much to lose
But young men think it is, and we were young.'

A. E. HOUSMAN

FOR AN HOUR we went on circling round, the pilot not daring to fly too low as another helicopter had already been shot down that day. It was a day no one was likely to forget, the 20th June 1967. From two thousand feet I looked down into the crater of a volcano, but only in the geographical sense was it extinct because the scene I watched through binoculars was one of fire and violence.

I was high above the streets of Crater Town, a name which stood for Aden in the language of South Arabia. What was happening there at that moment was the inevitable climax of weak British policy in Aden, where political expediency had for so long dominated military judgement that the result was this episode of needless sacrifice.

Although it was the siesta hour of an Arabian afternoon the streets below were crowded. This town of 80,000 inhabitants, the original city of Aden and its commercial heart, was shut off from the outside world by the steep volcanic cliffs which marked the crater of the dead volcano. The crowds I could see surging through the streets below were making their way eagerly towards two columns of black smoke and flame and a scattering of bodies spread across the road. These were the corpses of British soldiers. I knew that. And I knew that some of them must be men of my own advance party of the 1st Battalion Argyll and Sutherland Highlanders, who had died there so tragically beside their gallant comrades-in-arms of the 1st Battalion Royal Northumberland Fusiliers.

Of all the situations we had planned and trained for, this was a contingency I had never envisaged: a handful of Argylls committed outside my control. Yet even if I had been in a position to

1

intervene, most of the seven hundred Scots I commanded were nearly three thousand miles away in England, waiting to be flown out to join our advance party.

In five days' time Crater was due to become our area of operational responsibility, but on that day of 20th June I could only be an onlooker, my knowledge of events limited to scraps of information and what I could see with my own eyes. For that reason my personal account can never be complete and in any case it is less well informed than that of the Royal Northumberland Fusiliers whose conduct and performance during the time they were in Aden was, quite rightly, highly praised and widely recognized.

I had left England on 7th June with an advance party of one hundred and twenty-six officers and men. We had joined the Northumberland Fusiliers at their excellent barracks, named Waterloo Lines, which was on the outskirts of Aden and near the Royal Air Force base at Khormaksar. Here we quickly settled down to the well-tried drill with which the British Army hands over an operational area from one unit to another. Things went very smoothly and the Jocks—our nickname for soldiers in Scottish regiments—gradually took over the administrative chores while my company officers and other senior ranks were attached to their opposite numbers in the Fusiliers to learn the operational problems. I also had some officers learning Arabic at the army language school.

I was far from happy about the situation in Aden for reasons which will become apparent later in this story, but as the days wore on I became more and more concerned about the security of the Khormaksar area and indeed of Waterloo Lines itself. It may seem an old wives' tale but the fact that we were Celts seemed to bring us a sense of impending disaster. I could not pass it on to the others but all my senses were alive to the dangers of the situation rather than giving me any confidence in the way things were being managed. Then on the morning of the 19th the first of three revealing incidents took place.

I had a small office in Waterloo Lines, air-conditioned and a good spot in which to carry out administrative work. I was writing orders to cover our proposed deployment into Crater and got up to stretch my legs, idly looking out of the window. A staff car drove up to the side of the parade ground, behind the

2

operations room and about eighty yards away, and out of it got two senior Arab army officers, easily identifiable by the red bands on their hats. It looked to me as though they were reconnoitring the area and so I rang the guard room at the front gate to check who they were. By the time I got through they had gone but I asked the Fusiliers to see if they could get any more information. Their command post contacted the senior British staff officer with the South Arabian Army, getting the answer that these Arab officers must have been on their way to a meeting and had got lost. It did not satisfy me and my suspicions were confirmed when my Regimental Sergeant Major reported an exact repetition of the incident later that afternoon.

The same evening I had been invited to dine at Little Aden and the General Officer Commanding Middle East Forces, Major General P. T. Tower, was also present. Much of our discussion was over unrest in the South Arabian Army (SAA) which at that time numbered about eight thousand officers and soldiers. This had been brought about by the recent merger of the five battalions of the old Federal Regular Army with the four of the former Federal National Guard. It had upset the carefully maintained balance in the number of officers from the various up-country tribes from which the forces were recruited. In protest, eleven senior officers had presented a petition to their British commander, Brigadier Dye, and to the Federal Supreme Council, which was still trying at that time to control the newly-formed Federation of Aden and its hinterland. Four of these eleven Arab officers, all colonels, were therefore suspended from duty pending an investigation. From my own experience of serving with colonial forces in Africa prior to national independence I could see that the implications were very serious indeed.

I gathered later that a military intelligence report warned of the likelihood of serious trouble on the following day. Certainly it was grave enough for the GOC to have an additional escort, a half troop of armoured cars, across the causeway back into Aden that night after dinner. He was kind enough to give me a lift back to Waterloo Lines in his staff car and later I sat down and wrote a letter home to Sue, my wife:

Yesterday I had a pretty full and interesting morning in Crater with the people we are relieving. It still astounds me how the Government can get itself into a mess like this and make such little

3

effort to ensure that the soldiers who are supposed to sort it out have adequate powers. The Arabs are allowed to buy and carry arms and ammunition so that there is no way of knowing who is a terrorist and who is not. The Police are utterly powerless and, as the Army never gets any proper information, the situation is futile. I am convinced that once we take over in Crater there will be quite a shock coming for those in authority who refuse to face the facts. I am damned if any Argylls are going to get killed in support of useless policies.

Next morning, the fateful 20th June, I awoke early and continued the preparation of orders for our deployment. I had already decided that the company commanders had learnt most of what they needed from the Fusiliers and it was now only a question of briefing them on the new methods I was going to employ in Crater. New units invariably bring fresh ideas so this was no reflection on the Fusiliers. I had discussed with their Commanding Officer, Lieutenant Colonel Dick Blenkinsop, the possibilities of an alternative site to the Armed Police Barracks where the Fusiliers based themselves whenever they deployed in force into Crater. I did not like it as it was overlooked so I had started negotiations to requisition two alternative sites—one of these being a prominent house in a dominating position on the line of the old Turkish fortifications which surrounded the town. Blenkinsop had experienced nine months in Crater and his Battalion had been highly successful in putting down civil disturbances earlier in the year.

Just before noon I walked across to the officers mess in Waterloo Lines. As we were all now busily preparing for the imminent arrival of the Battalion I had called a meeting for that afternoon to give out my orders to the company commanders. Therefore I was surprised to see the tall figure of Major Bryan Malcolm, who commanded 'D' Company of the Argylls, striding purposefully across the patio of the mess. He called over to me in his usual cheerful manner: 'Sir, something's happened and they are going out. I'd better go too.' I nodded and he gave a great grin, pointing to his Argyll cap badge which I had ordered the Battalion to start wearing again after several years of being made to wear the badge of the Highland Brigade. He waved and went through the door. I never saw him alive again.

The fact that 'something's happened' surprised nobody in

Aden where terrorist attacks were a daily occurrence. But then I was approached by Lieutenant Brian Baty, my Reconnaissance Platoon commander, who said he would like to go too. I felt uneasy and there was a tension in the air which I could not explain, so I simply said 'No'. It probably saved his life. I sat around waiting for any news, had a bite of lunch and went over to the operations room to see what Dick Blenkinsop was up to.

Bryan Malcolm had meanwhile got into his Land Rover with two of our Jocks, Privates 'Pocus' Hunter and Johnny Moores of the Reconnaissance Platoon, and was following Major John Moncur, his opposite number in the Fusiliers, in another Land Rover with their sergeant-major and five Fusiliers, out of Waterloo Lines and towards Crater. At the gates of the cantonment, they saw my Battalion Intelligence Officer, Lieutenant David Thomson, driving back from a shopping expedition in Tawahi, the seaport area of Aden where the passengers from passing lines used to shop. They jokingly pulled his leg about this, repeating that 'something had happened' and that a 'State Red' alert had just been declared.

The background to the 'State Red' was the prevailing military philosophy that the best way to keep the peace in Crater was to keep out of Crater. Headquarters Aden Brigade considered that the presence of soldiers acted as a provocation to the militant nationalists and that for the Northumberland Fusiliers to show themselves too often in this highly volatile place would only result in unnecessary loss of life and injury to both sides. I did not agree. I thought this was a policy of appeasement. Appeasement always fails. The methods used by the terrorists in Crater stemmed from their having the initiative and were based on the 'come into my parlour' technique. Because the civil law was unenforceable every petty gunman or grenadier could wait for a suitable target and be almost certain of success in making his escape. Only a few nights before an off-duty party of sergeants in the Royal Air Force and Army had been brutally murdered whilst sea-fishing near the Supreme Court at the entrance to Crater. Equally significant, if one can compare murders, was the fact that a British ex-Superintendent of the Aden Armed Police had been murdered inside Crater within twenty yards of the front gate of the Armed Police Barracks.

I was personally enraged by these murders and the apparent

5

apathy with which they were accepted by the British authorities. Nothing, absolutely nothing, seemed to be done about them. Equally encouraging to the terrorists was that the security forces were never in sufficient strength to saturate the town of Crater with observation and fire, nor were there any properly maintained obstacles to movement between the narrow streets, on the Berlin Wall principle. I appreciated that the Aden Brigade tried to maintain a balance between offensive action, the need for rest and a reluctance to accept casualties. But, in my opinion, this had been achieved at the expense of a true military domination of Crater, though with the rules whereby Arab civilians were allowed to carry arms, and with a politically subverted police force, it was debatable how far a proper military domination could be maintained—even if it could be achieved.

Shortly after I arrived with the advance party the leading British bankers closed the doors of their main offices in Crater in protest against the lack of security for British employers. They threatened to pull out of Aden altogether. But not only were the British in danger of their lives. Hundreds of Adenis had fled in fear, many to the Yemen and a wealthy few to Britain. Protection money, assassination, kidnapping, torture and intimidation were a common part of life in this British colony. I considered that the Aden Brigade policy was 'containment' in the classic manner, that is, it was neither truly offensive nor purely defensive—middle of the road stuff. But confident that the Argylls were tough, trained and full of fight, I was planning a vigorous, military domination of Crater.

Because of the Aden Brigade tactics, the Fusiliers lived outside Crater in Waterloo Lines and, for much of their stay in Aden, kept no one permanently in Crater at all, relying on the Aden Police and the Armed Police, who had a large barracks in Crater, to keep order. The Fusiliers were divided into three main forces: an Internal Security operations company, which was responsible for Crater; a Guard company for protecting Waterloo Lines and military establishments outside Crater, and a Reserve company.

In the normal course of events a single platoon was responsible for the control of Crater by day. If trouble started this was reinforced by the remainder of the I.S. company. When a major disturbance was expected the Battalion tactical headquarters

6

was set up in the Armed Police Barracks and two companies were deployed in or just outside the town. At night any normal patrolling was done by a detachment operating from armoured vehicles.

It was my opinion that the terrorists' reaction to this was to try to lure the Fusiliers into an ambush. Crowds would start demonstrating in the streets and lighting bonfires of rubber tyres and giving the impression of a major disturbance in the hope that a patrol would be sent to investigate so that the narrow streets along which they would have to pass could be ambushed by snipers and grenadiers. There had been some ugly rioting in Crater and the Fusiliers had bravely gone in to grapple with mobs who heavily outnumbered them. It says much for their skill that, during their nine months in Aden, they had, until 20th June, not lost one man killed.

So, when I reached the Fusiliers' Operations Room, I thought I knew what to expect. Dick Blenkinsop was there and he told me that a report had come in from Crater that soldiers had been wounded and were pinned down by enemy fire under the cover of the outer wall of the Armed Police Barracks.

Blenkinsop was talking on the radio to his I.S. company which was standing by at the seaward Marine Drive end of Crater, trying to discover what had happened. I asked whether any of my people were involved and he said that he did not know. I stood around like a spare bridegroom at a wedding, filled with foreboding. Later he managed to get through on the telephone to Superintendent Mohammed Ibrahim who commanded the Armed Police, asked him what had happened and sought his help in evacuating the British soldiers who were then thought to be wounded and trapped. Ibrahim apparently said that he could do nothing, that there was a lot of firing but he would do what he could to stop this. The fog of war was thick in that Operations Room and I felt extremely sorry for Blenkinsop.

By now I was as confused as everyone else as to what was happening in Crater but I was determined to find out for myself as I was worried about the safety of Bryan Malcolm and his two Jocks, coupled with obvious concern for the Fusiliers and interest in a situation that I was on the eve of inheriting. I assumed that Bryan Malcolm must have followed John Moncur into Crater, for it seemed that two Land Rovers had been involved

7

in whatever fighting had taken place. I did a bit of badgering and at three o'clock, three hours after the first reports of trouble in Crater had filtered through, a Scout helicopter was made available to me. I grabbed David Thomson and we got in and took off for the dark line of cliffs which hid Crater from the outside world.

A helicopter carrying part of the Northumberland Fusiliers' Intelligence Section had made a brief reconnaissance over Crater shortly before and an earlier helicopter, carrying Fusiliers to set up an observation post on the crest of the cliffs above Crater, had been shot down by the enemy at midday, luckily without loss of life. As far as I know David Thomson and I were the only two officers to go and see for ourselves what was happening. As the senior British officer in or above Crater, I was without authority to give any commands, nor indeed had I been asked to express any views. I was a 'New Boy' to those experienced Aden campaigners who consistently and mutually confirmed their own views on the proper conduct of internal security operations in a British Colony.

As the helicopter climbed through the hot afternoon air to clear the cliffs I gripped airman's binoculars with rubber eye-pieces and focused them on the scene below. We looked down on to the flat roofs of Crater Town with its grid-pattern of streets, opening out into market places, the minarets of its mosques, the squalid shanty-towns that climbed up the volcanic slopes and onto Queen Arwa Road. At the point where this road passed the high wall of the Armed Police Barracks, David and I saw what we both knew, but never admitted to each other, we were looking for.

Two Land Rovers lay burning in the road and around them sprawled dead bodies. If they had been ambushed the shooting could only have come from the Armed Police Barracks and a row of high, modern, concrete flats on the opposite side of the road. Crowds milled and surged in the streets below, sometimes suddenly scattering as though being fired upon. Smoke arose from several parts of the town; some from burning road-blocks and heaps of tyres—and others probably from Jewish shops, which we knew were being set alight in revenge for the defeat of the Arab armies by Israel.

Circling at two thousand feet—just beyond the accurate range

8

of small arms fire—we looked down on Crater for an hour. Had I been in a position of command I could, from my aerial command post, have ordered the immediate mounting of a rescue operation into Crater. Before taking off I had formed my advance party into a scratch fighting company in the hope that this might be ordered. But no such order was given. That this simple and straightforward order was denied was the most astonishing, and, to my mind, disgraceful act by the higher command. As far as I was aware no senior commander had actually seen for himself what was happening in Crater—it was as though it was all another telephone battle like those at the Staff College at Camberley or a training exercise in Germany. No order to go in and seize the crisis by the throat was given that day, and when dusk fell it was already too late.

Crater was firmly in the hands of the enemy. My sympathy went out to Blenkinsop, on the eve of his departure from Aden after such a successful tour of duty, faced with as nasty a situation as any commanding officer had seen. But by the present rules of operating he was bound to his Operations Room, a victim of weak military policy.

The full official report of what is supposed to have actually occurred on 20th June was never shown to me; perhaps much of what had happened will never be known because the men concerned are dead. From the time three days later, when I took over responsibility for the Crater area of Aden, I could vouch for the truth of events. But on the 20th June 1967 I was still a helpless and angry spectator.

It seemed incredible later, but throughout the morning of 20th June I believe that most British Army officers in Aden had not known that an eruption had already begun. It had started early at Lake Lines, only a few miles from us at Waterloo Lines, when young Arab soldiers at their Apprentice School, incensed by the suspension of the four colonels and tribal rivalries, had mutinied and set fire to barrack buildings.

Chain reaction was swift. Although officers of the South Arabian Army quickly dowsed the first outbreak and the four colonels were immediately reinstated, trouble spread across to Champion Lines, the Headquarters and Training Depot of the former Federal National Guard. Arabs broke open the armoury and started fighting each other in tribal groups. But this was not

9

all. Once discipline had snapped they went berserk, firing out of their camp at Khomaksar airfield and at Radfan Camp, where two other British battalions were based, the 1st Parachute Regiment and the Lancashire Regiment. At the height of the trouble the Arabs most cruelly ambushed a British Army truck carrying soldiers of the Royal Corps of Transport who were returning from practice on the rifle range. These British soldiers suffered a terrible fate, pinned down without any succour, and eight of them were killed. Here the situation was out of control and a company of the King's Own Royal Border Regiment from Bahrein, attached to the 1st Parachute Regiment, had to be called in to restore order, suffering casualties in the process. But by this time garbled accounts of the mutiny and subsequent fighting had travelled from the flat, dusty isthmus, where it had taken place, over the mountain barrier into Crater. And now Crater itself was to erupt.

The background to these events was the war between Israel and the Arabs, which had broken out only a fortnight before. The Arabs of Aden were nervous and on edge. They saw themselves humiliated before the eyes of the world and they were frustrated, furious and frightened. It was an explosive mixture of emotion which a spark could ignite, or a firm lead extinguish. The potential force of such an explosion was particularly potent in Crater. Not only was it the central citadel of the two main terrorist organizations, the Front for the Liberation of South Yemen (FLOSY) and the National Liberation Front (NLF), but the Armed Police were known to have been deeply infiltrated by terrorist organizations.

Thus, at 11 a.m. on 20th June, when garbled reports of fighting between British and Arab troops outside Crater reached them, the Armed Police announced from their barracks beside Queen Arwa Road that they would shoot any British soldiers who approached. Crater was now nothing less than a death trap.

The 'State Red' alert took effect from the late morning. Meanwhile, Major Moncur's company was deployed at the Marine Drive entrance to Crater in readiness for any trouble in the town. Earlier that morning, at about 10 o'clock, Moncur had sent a patrol into Crater in a 'Pig', an armoured 3-ton truck of the type used during the Malayan emergency and other Internal Security contingencies, capable of carrying a section and pro-

10

tecting them from small arms fire. This patrol was commanded by Second-Lieutenant John Davis and was accompanied by one of my subalterns, preparing for our takeover from the Fusiliers, Second-Lieutenant Donald Campbell-Baldwin. It was Campbell-Baldwin's account of what happened together with my direct observation that led me to reconstruct the sequence of events.

Escorted by a Saracen Armoured Personnel Carrier of the Queen's Dragoon Guards, Davis's patrol had moved cautiously through the seaward end of Crater but found nothing unusual. The same patrol was sent out again, but this time the atmosphere was very different. Campbell-Baldwin later told me that they were despatched at short notice and it was not discovered until after they had left that the A41 type wireless in the Pig was not working. The streets were deserted but they found one or two minor road-blocks constructed of chairs and other furniture, all of which were easily knocked aside by the vehicle. There was now no contact with company headquarters back at Marine Drive. John Davis, the platoon commander, an excellent young officer held in high esteem both in his own battalion and by my advance party, decided to alter his route back and leave Crater by the Main Pass. But about 300 yards short of the Armed Police Barracks they found buses had been parked across Queen Arwa Road as a road-block. They dismounted and moved these aside before driving on, but immediately came upon a threatening scene. Around the Armed Police Barracks, Arab policemen, all armed, were taking up firing positions commanding the road, both on roof tops and behind built-up defensive positions, or sangars. But the patrol passed this obvious ambush position and continued to the top of Main Pass, the landward exit from Crater. There they heard two shots fired behind them.

It would seem that Lieutenant Davis now had the choice of returning by the same route, facing the real risk of being fired upon by the ambush he had seen, perhaps by anti-tank weapons as well as small arms, or of carrying on over the top of Main Pass and driving round the outside of the Crater mountains along Marine Drive, which would bring him back to company headquarters in about ten minutes. He chose the latter course because he probably appreciated that until he got his radio repaired his information could only be delivered verbally and this was the safest method of delivering a warning.

11

It seems, though, that Major John Moncur, perhaps worried by the lack of news from his patrol, had decided to go himself and see what had happened to it. He in his Land Rover, followed by Bryan Malcolm in his, with Privates Moores and Hunter, therefore set off along Queen Arwa Road towards the Armed Police Barracks.

By this time, Davis's patrol had gone on over Main Pass and Moncur passed the shifted road-blocks into the wide, straight boulevard. At the point where this passed the Armed Police Barracks, the ambush was sprung. From the barracks and the flats opposite, machine-guns, light automatic weapons and rifles opened fire. The Land Rovers pulled up and those men who had not already been hit leapt out to fire back. But there was no cover and the massacre was over quickly. Only one soldier, Fusilier John Storey, managed to race across the road to the cover of the flats without being hit. It was said that when he looked back, all the others appeared to be dead, only one man still firing from the meagre cover of his vehicle. Then both Land Rovers caught fire and another eight British soldiers lay dead to match the morning massacre of the Royal Corps of Transport.

Fusilier Storey got on to the roof of the flats and exchanged fire with some terrorists who spotted him there, then burst into a flat and hid for three hours, holding the occupants at gun point. Finally, trying to escape, he was captured by an Arab policeman and taken to the Armed Police Barracks, where he was apparently put into the back of a truck with the burned bodies of the British dead.

To discover what had happened to his company commander, the gallant Lieutenant Davis with Donald Campbell-Baldwin still in attendance, again took his patrol up Queen Arwa Road but still, unfortunately, with his wireless out of order. As they saw ahead the burning Land Rovers and the dead soldiers sprawled in the road, they themselves came under heavy fire. While the Armoured Personnel Carrier stood in the road return-fire with its machine-gun, the Pig pulled up behind the cover of a garage and the patrol dismounted. Davis and three Fusiliers climbed up a drainpipe and took a machine-gun into a nearby building with the intention of giving covering fire to any survivors of the ambush. He then ordered Campbell-Baldwin to return with the Pig and the Armoured Personnel Carrier to

12

company headquarters, report what had happened and get help.

When Campbell-Baldwin got back, the Support Platoon Commander of Y Company of the Fusiliers ordered another mobile patrol to set off down the road, again commanded by a Fusilier subaltern but with Second-Lieutenant Douglas and Corporal Currie of the Argylls, both from Bryan Malcolm's company, included in the numbers. With Campbell-Baldwin they constituted the only three Argyll survivors who were able to witness the events of that afternoon on the ground. This time the accompanying vehicle was a Saladin armoured car armed with a 76 mm. gun. It, too, came under heavy fire and, deciding that to stop, dismount and attempt to reach any survivors would be certain death, the patrol commander sped past the Armed Police Barracks under intense fire and left Crater by Main Pass. I gather that the troop commander had been refused permission over the wireless to fire anything heavier than his machine-gun. This was the crucial decision. In my view, as a soldier, it was the wrong decision.

At 4 p.m., telephone contact was again made with the Armed Police Barracks and an amnesty arranged to allow an ambulance to collect the wounded and any survivors. Although it was fired upon, this ambulance got through, but only one survivor, Storey, was handed over. As I was never shown the proceedings of any official inquiry into the Armed Police mutiny, nor asked to give evidence, I could only assume that at that time Davis and his three gallant men were still alive. The uncertainty of their fate has never been satisfactorily explained to me but the manifest horror of it became such a strong factor in my subsequent attitude in the months ahead, when my own men were deployed in isolated section posts throughout Crater, that I would hesitate to comment on the manner in which they met their end. What is certain is that the circumstances of their being abandoned confirmed my complete lack of confidence in the military management of Aden, though it served as an example of devotion under fire which is in the highest tradition of the British Army.

Crater was now in the hands of the enemy. This fantastic, nightmarish situation, where in the middle of a British Colony in peacetime the rule of the Queen is abandoned, had actually taken place. And worse, it was about to be condoned by Her Majesty's Government with 22 dead British soldiers to prove it.

13

Crater was held by at least 500 well-armed mutineers of the Armed Police and terrorist gangs. Objective sources estimated that the re-capture of Crater would require a brigade attack, involving heavy military and civilian casualties.

The arbitration with the Armed Police to recover the bodies of the dead was charged with drama. Was a police vehicle booby-trapped? Why did they leave it abandoned outside the Roman Catholic Mission in Crater on the evening of the 20th June for a period of 24 hours? Even the final macabre process of identifying the bodies in the early hours of 22nd June left some of my questions unanswered—though it confirmed others.

Why had this tragedy happened? How was a bad situation allowed to get worse? When it became obvious that Davis and his men and any survivors of the ambush could not be rescued without using the Saladin armoured car's gun, why was this not allowed? This gun was capable of blasting the enemy from behind their cover of sandbags, sangars and concrete balconies. And finally, why was a straightforward infantry assault not authorized?

Crater was now firmly in the hands of an enemy who had inflicted a humiliating defeat on the British Army, killing twelve men of the twenty-two dead that day—including two company commanders of high promise—and wounding others. Expecting the immediate counter-attack that did not come, the terrorists at once consolidated and strengthened their positions. All British troops having been driven from Crater, the road-blocks that were set up on the two entrance roads and the observation posts set up on the surrounding mountains, came under intensive fire.

The enemy was understandably jubilant. All the prisoners in the Jail were released, to add their weight to the terrorists. In the eyes of the Argylls this was a disgrace.

I had no sympathy to waste on those in authority whose weak policies had led to this intolerable situation. I did feel extremely sorry for Dick Blenkinsop and the Northumberland Fusiliers. They had been ordered not to go into Crater again because of the political consequences of a battle against Arabs. The irony of it was that this same Federal Government which Whitehall was apparently so anxious to support collapsed some few months later, before our evacuation. Therefore the events of the 20th June 1967 in Crater were a classic example of military judgement

14

being influenced by political expediency. I state this advisedly in view of the clever arguments used to justify those in authority who took the decision to abandon Crater.

Two days after these terrible events, the main body of my Battalion began to arrive and, on the 25th, I took over responsibility for Crater from Dick Blenkinsop. I did not like the idea of the British Army being so humiliatingly defeated, particularly by so third-rate an enemy. I made clear to all that my intention was to re-capture Crater at the earliest possible moment.

Others, senior to me both in Aden and Whitehall, were not so eager to re-establish our authority. If a brigade attack, supported by heavy weapons, was made on Crater, they said, the result would not only be heavy civilian casualties but the assault force might suffer thirty per cent losses and in the crowded tightly packed houses and flats of Crater Town, hundreds of civilians might die. As a result of this, the South Arabian Army might mutiny and British troops outside Crater itself could have a major battle on their hands. Also, the two hundred or more British civilians and Servicemen working in the hinterland might be massacred. Inaction was supported by fear. The official line, frequently expressed, was that 'we must play it cool'. But by playing it cool on 20th June, we had found ourselves in our present disgraceful predicament.

In London it was being forecast that the British Army would never be able to re-take Crater and that Nationalist flags would continue to fly from its rooftops and minarets until we were finally evacuated or driven into the sea. Even British correspondents in Aden itself reflected the official view. The *Daily Telegraph* correspondent, Christopher Munnion, wrote: 'British troops may never return to patrolling the Crater area of Aden, now declared an "independent republic" by nationalist extremist groups. . . .' And in the *Observer* John de St. Jorre wrote: 'There now seems little doubt that British troops will never re-enter the "secessionist" Crater district of Aden.'

The extent of the British humiliation was made clear in a communique issued by the FLOSY terrorist movement, the facts of which were indisputable, and were broadly that 'the Revolutionaries defeated the enemy and made them retreat'. The communique concluded: 'FLOSY is the leader of the sacred national struggle and has made heroic sacrifices in all the battles

15

which it has led against the Occupation Forces. A number of its men have fallen in the cause of duty and right and dozens of its members have been detained. FLOSY will not weep for them because all of us are pledged to this homeland and we are all prepared to make sacrifices to liberate the homeland whatever the price. . . .'

For the terrorists in command of Crater, the sight of British transport aircraft had become a familiar sight. But what they did not know was that many of the aircraft circling above them during the first few days of their rule in Crater were carrying men trained and toughened and prepared to re-capture their seemingly impregnable stronghold. The Argyll and Sutherland Highlanders were coming.

For me it was the first day of my twenty-fifth year in the British Army. As I waited at Khomaksar for the Battalion to arrive my mind went back to how it had all begun. Now, with this situation facing us, I wondered just where it was all going to end.

CHAPTER 2

'Here lies Kartoffelstein
Latterly called Fitzwarren
There is some corner of an English Field
That is forever foreign.'

JAMES AGATE

EVERY SPRING MY mother used to unpack an old Army uniform dating from the First World War and hang it on the washing line in our small back garden to 'give it an airing'. The sight of my father's khaki tunic, with its lieutenant's badges of rank, three gold wound stripes and the faded white and purple ribbon of the Military Cross, together with the kilt and sporran of the Argyll and Sutherland Highlanders, was my first introduction to the Regiment which was to become, for me, a second family. Goodness knows how old I was when I first remember the sight but it fills a portion of my early memories, a kaleidoscope of scenes—the ill-fated airship R101 flying over our home, the Crystal Palace burning down, the Horse Show at Olympia, catching the Flying Scotsman at Euston, and all the visual impressions which came the way of a boy born on 17th November 1925 in London S.W.16.

The garden lay behind a small, semi-detached house in the suburb of Norbury. The setting was typically English suburban middle class and I was to grow up in that sensible and happy environment. But at a fairly early age I began to realize that although we lived among English people there was something that made us different. My sister, Hetty, five years older than me and always a thoroughly helpful and friendly person, used to explain this in a very matter of fact way in the clear English accent of Selhurst Grammar School for Girls: 'We're Scots, of course'. This apparently explained everything; certainly to her satisfaction if not mine. I used to think a lot about it and it seemed to tie in with my second christian name, Campbell, which was at such variance with all the local Peters, Davids,

17

Freds, Bills and Toms. In fact for a long time I thought Campbell was an original name and was rather shy of it until I realized there were a few more in Scotland and Sir Malcolm Campbell started racing Bluebird—whereupon forgetful old ladies always called me Malcolm, which was even more confusing.

Years later someone told me that I had in my veins the same mixture of blood that made the Argyll and Sutherland Highlanders unique—being born with the strong emotions and romanticism of the Celtic Highlander and the tough practicality of the Scottish Lowlander. My father, after whom I was named Colin to add to my Campbell, was a West Highlander, and my mother was from Glasgow. These two strains in my blood were bound together by powerful Scottish patriotism. Distant as we were from our true home, Scotland was never far from our thoughts and lives. It was to be the same in the Argylls, where long years of overseas service never weakened the link with home.

But in the early 1930s these profound thoughts certainly did not occur to me, as my sister taught me how to ride a bicycle on the forbidden fairways of the private golf course, long turned into a housing estate, which backed onto our garden. At the local municipal baths, she also taught me how to swim, and I used to go off, a tiny hanger-on, with her pack of older girls and boys who lived in the neighbouring houses. Picnics in Norwood Grove; rides to Regent's Park Zoo on the 159 bus; hours of fascinating tours round the museums in South Kensington; trips to Banstead Common and Epsom Downs and the world of Rovers, Scouts, Guides and Cubs which circulated around our local Parish Church, St. Oswald's. From Hetty—or Penny as we came to call her—I learnt that enthusiasm and enterprise go together.

My paternal grandfather and his forebears came from that part of Argyll which was first colonized by the Dalriadic Scots, who crossed the North Channel from their home in Antrim to Kintyre and drove out the Picts in A.D. 258. These were the men who made a raid through Roman Britain which carried them to London before they were repulsed by Theodosius. But tracing their origin and kinship is an impossible task for a layman. All I know, from what I was told and have seen on the gravestones in the little cemetery at Lochgilphead, the village where my

18

father himself was born, is that for the last few generations my forebears had pursued the normal rural life of Argyllshire village folk on the shores of Loch Fyne as fishermen, farm workers, shepherds and gardeners, while many of the women became nurses and were figuring as matrons of hospitals by the turn of this century. They married locally—my great grandmother, Elizabeth Campbell, coming from Minard further down the loch on the road to Inveraray.

My mother was a Gilmour from Cathcart in Glasgow. Her own mother was a Bowie, an old race of agriculturalists who had farmed in the Paisley district of Renfrewshire for many generations and whose name is found in South Africa and Canada while famed in America at the Alamo. My maternal grandfather, John Gilmour, was the carting superintendent of the old London, Midland and Scottish Railway and one of my earlier thrills was being taken round the long stables of Clydesdale horses which in those days plodded majestically through the cobbled streets of Glasgow pulling the freight and merchandise to and from the railway yards, docks and warehouses. He was a great judge of horses and in Scotland, at agricultural shows and exhibitions, his bowler-hatted figure would be prominent in the show ring.

My own father was educated in the village school at Lochgilphead, setting out to work locally and then in Glasgow where so many young country boys of his age and background went to seek their fortune. He worked for a solicitor and later for MacBrayne's shipping company which linked the West of Scotland by boat and steamer. When the First World War broke out he joined up, like most of his generation, and enlisting in Glasgow was sworn in as a private in the Highland Light Infantry. He used to tell me when I was a boy that when they mustered at Maryhill Barracks there were no uniforms and no rifles so they were issued with Glasgow Corporation tram conductor's uniforms and armed with broomsticks. He also used to joke that the British Army promoted by size and this had happened to him. I suspect that it was probably his natural authority that picked him out, or was it because he was a keen footballer? Anyway, he was promoted and went home to Lochgilphead on his first leave as a Sergeant. In the village street he met the local laird, Malcolm of Poltalloch, who said, 'You should be an

19

officer in our own county regiment, the Argyll and Sutherland Highlanders.' And so, in 1915, my father was commissioned in the Regiment and later joined its 10th Battalion in France, part of Ian Hay's immortal 'First Hundred Thousand'.

Like all small boys born within ten years of the First World War, the thought of it was constantly with me and I was always asking my father to tell me what had happened and about his own adventures. He was always reticent. Probably he was so amazed and thankful to have survived that he just did not want to be reminded of it. I knew that he had fought in most of the big battles and had been wounded three times and, on the last occasion, in 1918, had been hit and gassed so badly that he never returned to France. I knew also that he had won the Military Cross at the Battle of Ypres but when I asked him how, he would only say, 'Oh, shooting rabbits.' But my mother showed me the citation and I still treasure the gold watch chain which was presented to him by the people of Lochgilphead, for he was their local boy who had achieved distinction.

While convalescing from wounds with the 3rd Argylls at Kinsale in Southern Ireland he had met my mother and they were married when the war ended, He returned to MacBrayne's as a ship's purser. After the birth of my sister, they moved to London in 1921 and eventually bought the little house in Norbury. My parents brought to England with them their Scottish accents and customs, but my father had a great admiration for Londoners and thought Cockneys the salt of the earth. His choice of light entertainment put Flanagan and Allen at the top of the bill—so did mine. We spent many of our holidays in Scotland and I was always brought up to be intensely proud of my ancestry and its traditions. When I was only six, my father would, once a week, invite a friend of his who was in the London Scottish Territorial Army unit to the house to attempt to teach me to play the bagpipes and Highland dancing. I can still picture myself, dressed in the kilt, dancing a Highland fling in our small sitting room in Norbury, the sound of the chanter carrying out across the suburban lawns, rose trees and privet hedges.

I wore the kilt on Sundays, too, when I was taken to the Presbyterian Church in West Croydon, where my father was an Elder. A man of stern integrity, my father brought me up according to his own strict moral code. He had a strong sense of civic

20

responsibility and, for example, if a postcard arrived from the public library saying that a book we had borrowed had been kept too long, he would launch into a sermon about our duty to our fellow-citizens who might have been waiting to read that book. A small failing such as this would be regarded by him as a sign of irresponsibility.

Although gentle he could be deeply stirred at what he considered failure in others to do their duty. Until about 1938, for example, he thought little of Winston Churchill because he suspected he had broken his parole when a prisoner of the Boers. And one of the rare arguments I had with him was at the time of the abdication of King Edward VIII. I had ventured to say a word in favour of the ex-King to find that my father was shocked and appalled that a son of his was taking the part of a man who had abandoned his duty.

Our patriotism was not solely Scottish. We felt immense pride in the British Empire and believed with a self-confidence that was almost Roman, that it was entirely right that the British should rule a large part of the world.

I was first sent to school for several tough but stimulating years at the junior part of Streatham Grammar School. This entailed a bus or a bicycle ride with a shilling in my pocket to get lunch. My mother was a tremendous person for making us stand on our own feet but at the same time offered good advice and a refuge if things went badly. She let me choose my daily restaurant. It was usually either the Lyons Tea Shop at the top of Streatham Hill where 1s. went a long way in 1935, or a rather grander café where you could still get a 'schoolboy's lunch' for 9d. and leave 3d. over for sweets.

Meanwhile I had joined the St. Oswald's Church Boy Scout Troop in Norbury and all my spare time was devoted to the exciting world of Scouting. We had tremendous fun and companionship. With our trek cart and rucksacks we went far afield. Along miles of country roads we rarely met a motor car, train fares were cheap, hostels gave you food for a song and if it rained you could always play Monopoly in some friendly farmer's barn. Once in Gloucestershire, we were invited in to meet the local landowner in whose park we were camped. It was dark and we walked up to what I thought was a vast palace of a house. We were taken in to a 'Den' and there the owner, a

21

retired Army Colonel, sat surrounded by the most fascinating collection of relics I had ever seen. The heads of wild animals snarled from the walls between native spears, knives, rifles and odd-shaped masks. He told us of shikari in India, of safari in Africa, and of all the adventures of colonial soldiering. 'That', said I to myself, 'is for me.'

Our family fortunes were now prospering. My father was doing well in the marine insurance branch of a London solicitors and his utter integrity gave him increasing influence and standing. In 1937, when I was twelve, we bought a larger house in Purley and I went on to Whitgift School, a large, day public school at South Croydon, where I was able to transfer my enthusiasm for Scouting to the Officers' Training Corps, in which I eventually became an under-officer.

In school my best subjects were English, History and Theology —but, outside school hours, I continued to read avidly. I read the Classics, modern novelists and the popular adventure books of the day by John Buchan, Dornford Yates, and 'Sapper'. I particularly enjoyed historical biographies but the book that had the most profound influence on me was *Seven Pillars of Wisdom* by T. E. Lawrence. I understood about a quarter of what he wrote but recognized that this was what life was surely all about—great causes and lost hopes; war and philosophy; men of action combining intellect and genius with physical endeavour; romantic foreign places where life was cheap but honour abounded. It fascinated me and it baffled me. But Lawrence of Arabia became a hero and very much a prototype of my future heroes: a man of action who thought for himself and acted on his own convictions.

I also read the lives of soldiers and sailors and accounts of their campaigns. I was more interested in the lesser-known but more individual soldiers—such as Probyn and Hodson in India —than in the great military heroes of the nineteenth century— who struck me as being rather pompous except for Field Marshal Roberts. This interest was fully supported by my father, who was a great admirer of the higher Victorian virtues although rather critical of Queen Victoria because of some long-held prejudice about John Brown, her Highland ghillie.

Even before we moved to Purley, it seemed inevitable to us all—even to myself as a small boy—that war was coming.

22

Mussolini I thought of as a greater villain than Hitler because of his attack on Ethiopia and the bombing and gassing of the tribesmen and by his terrible execution of the rebellious Senussi chiefs in Libya by having them thrown out of aircraft. When war did break out in 1939, I can remember walking with my father to the station one morning, and complaining that it would all be over before I was old enough to fight. He told me that I had no need to worry about that: the Second World War would, he said, last as long as the first.

Purley lay not only near Croydon, the main London airport of the time, but also near the Royal Air Force fighter stations of Biggin Hill and Kenley. Thus, when the storm broke in the summer of 1940, it broke over our heads. I remember walking with my sister one fine, sunny day and she pointed up and said, 'Look at all those aeroplanes!' Now, like every schoolboy, I was expert at aircraft recognition and I promptly announced that they were Junker 87 dive bombers heading for Croydon. As I spoke, the bombers turned on their noses and dived down to drop their bombs on the airfield below.

Dog-fights between Fighter Command and the Luftwaffe swirled above our heads and the blue sky was streaked with the long curves and twists of white contrails. Our garden was often littered with spent cartridge cases and fragments of anti-aircraft shells. We listened avidly to the evening 'score' on the BBC and like every young Briton I hero-worshipped the RAF fighter pilots. Air Marshal Tedder and Group Captain John Cunningham were both old Whitgiftians so there was always a strong RAF 'lobby' at school.

Like most boys of my age, I regarded the war as a thrilling adventure and longed to take an active part. We children soon became involved. That winter of 1940 the night raids began and fire-bombs rained down on British cities. One night, in particular, loads of incendiaries showered on to Purley and we all rushed out with buckets of sand and shovels to put them out and to douse fires with stirrup pumps. We had, early on, constructed quite an elaborate shelter in the orchard at the bottom of our garden but the raids became so continuous that we grew sick of trailing to and fro, from the house and over the terrace, rockery and lawn to the shelter and back. So my father had a more elaborate 'dug-out' built under the garage. But while it

23

was being constructed we all got under the billiards table with its thick slate top and sat on cushions, chatting and listening to the news and comedy programmes—Tommy Handley and Arthur Askey the comedians of the day. Although one high explosive bomb destroyed a house in the road adjoining we were always calm and quite naturally so. If we had needed a calming influence, my father would have provided it with his relaxed, knowledgeable comment on each explosion, remarking on the size of the bomb and its distance from ourselves with an occasional nervous clearing of the throat as the only indication that he might be worried. I thought it was all quite fascinating; so did our dog.

My father joined the Local Defence Volunteers, later renamed the Home Guard, the day Anthony Eden called for volunteers. I longed to join, too, but the lower age limit was sixteen and a half and I was only fifteen, so I had to content myself with cleaning my father's rifle and equipment and boots. But the men of 6 Platoon, B Company of the 58th Battalion, Surrey Home Guard, were mostly middle-aged or younger men waiting to join the Services. They needed a runner who could act as messenger on foot or bicycle and I begged them to allow me to do this.

So now began my first insight into the strange mind of military officialdom. It was requested that I be enlisted in the Home Guard, although under age, for operational reasons. After initial refusal, the War Office agreed, on condition that my father signed a form of indemnity so that, should I be killed or wounded, no claim for compensation could be made on them until I reached the age of sixteen and a half. So, at the age of fifteen, I proudly put on my uniform as probably the youngest Home Guard in Britain. My father thought it was quite right and proper that I should join him so we became 'Old Mitch' and 'Young Mitch'. I still cleaned the kit for both of us. My father thought the best thing was the War Office concern about indemnity, 'Typical of the civil service mentality', he would say in his West Highland exactness. 'I suppose they think the Germans will take advice from the Treasury Solicitor before they shoot at us both.' But looking back at it all with the hindsight of experience and as the father of three children myself, I think it was a brave effort on my parents' part to accept my enthusi-

24

asm to share the danger because they were by no means unimaginative and my father knew what war could be.

We were sure that the Germans would try to invade and we knew, quite rightly, that, if they did, one of their main objectives would be Croydon airport. In my imagination I saw the Panzer corps roaring through the streets of Purley with nothing to stop them but my father, myself and the rest of 6 Platoon. We therefore rehearsed our plan to defeat the Wehrmacht, just as thousands of other little groups of British middle-aged and elderly men and boys were doing all over the United Kingdom.

Our Tank Destruction Team consisted of six men. They would go into action on the following theoretical drill. A German tank would be assumed to be approaching down the street, whereupon Number One would step out from behind a privet hedge and stop it: how he would stop it I cannot exactly remember, but I think by holding up his hand, as if he had a message or warning for the tank commander. Once the tank had stopped, Number Two and Three ran from behind the hedges and rose bushes, on either side of the street, carrying large bundles of hay which they flung on top of it. At this moment Numbers Four and Five appeared carrying a bucket of petrol and a stirrup pump with which they sprayed the hay. This completed, all five would take cover and Number Six would appear to throw a grenade at the tank and ignite the hay thus, we confidently expected, destroying the tank and all within and halting the German advance on Croydon.

Otherwise we were kept busy with guard duties and night patrols and sometimes we went over to the Ranges belonging to the Guards Depot at Caterham to be instructed in the use of such strange mortar-like weapons as the Blacker Bombard, the Spigot Mortar and the Northover Projector, which were widely held to be more lethal to their crews than to the enemy.

My sister now joined the ATS—on the tacit understanding that she should not go into the more fashionable WRNS, because we were 'an Army family'. I longed to join up myself, but the lower age limit for enlistment in the Army was seventeen and a half. So, impatiently, I carried on at school, putting my warlike instincts into Rugby football and becoming captain of swimming. Much of our time at Whitgift was spent in the air raid shelter and before playing Rugger we used to form a line to clear the field of odd bits of shrapnel. At this time, my father thought that I should begin to prepare for my post-war career which he saw to be in the Law. I accepted that my aim should be to read for the Bar. I never attained the Sixth Form, although I became a prefect, because I remained in a class called 'Middle Services' among boys of brawn and bravado whose only interest was, like my own, to go to war at the first possible moment.

Finally, I could wait no longer, and, on 17th May 1943, I walked out of school and into the Croydon recruiting office and there enlisted in the Army. That evening, at supper, I announced that I had joined up and would go in a month or two, as soon as I had reached the age of seventeen and a half. My mother took it very well indeed and my father at once gave me some practical advice. Naturally I wanted to become an officer in the Argyll and Sutherland Highlanders, he said. But it was inadvisable to join as a private soldier the regiment in which you hoped to be commissioned. Thus, on his advice, I joined an English county regiment, the Royal West Kents, but all my future aims and ambitions were centred upon one goal: a commission in the Argyll and Sutherland Highlanders.

A few weeks later, my call-up papers arrived and, on 1st July, I walked into Maidstone Barracks with my suitcase to become 14432057 Private Mitchell, C. C. in the General Service Corps.

We were the most astonishing collection of new recruits, ranging from a tough old navvy who insisted on keeping his trousers up with an officer's Sam Browne belt, to a pilot who had been dismissed from the Royal Air Force for 'lack of moral fibre'. We also included a good number of policemen. They were leavened by two characters who had been in prison, one chap who had fought against Franco in the Spanish civil war, a ballet dancer and a boy who had returned from America as an early war-time evacuee. I suppose that we were typical of hundreds of similar squads all over the country.

We were drilled, inoculated, indoctrinated with 'British Way and Purpose', taught to fire weapons and generally converted from pyjama-wearing civilians into rough-shirted basic soldiers. I did not find it very difficult and after six weeks moved from the old Barracks beside Maidstone Prison up to the Infantry Training Centre at Invicta Lines. Here began six weeks of more advanced infantry training at the end of which I was put on an

26

NCO's course for another six weeks and promoted lance corporal. My sister, now a subaltern in the ATS, came down to see me and we marched around in step returning each other's salutes after which I bought her tea in the Star Hotel and called it a day.

Being the only lance corporal who instructed on the NCO's cadre (the others were all sergeants), I became a sort of 'odd-man-in' and lived in the gymnasium with the physical training staff who were also mostly corporals. They were fitness fanatics and spent most evenings practising on the wall bars. The Staff Sergeant Instructor was Stan Cullis, the soccer-playing Captain of England, and I felt like a fly on the wall of Olympus to be living so close to the great man.

We worked hard and I rarely went out at night, reserving my relaxation for Saturday afternoon when I changed a book at Boots' Library, went to the cinema, had an egg and chip supper in the YMCA and went back to clean up for Church Parade. Occasionally we got a 48-hour pass when we all scrambled into the overcrowded train to London. I would then go out to Purley and spend my time playing golf and generally lazing around while my mother very sweetly took charge of the inevitable haversack full of dirty washing that every war-time soldier seemed to carry back to his wife, mother or possibly girl friend, if he was lucky enough to have one out of the Services or a munitions factory. I kept steadily going with a Wren who was billeted in Chelsea and whom I had known for years. I always enjoyed female company but had a sort of suburban teenager's interpretation of Napoleon's great dictum, 'Love is the occupation of the idle man, the distraction of the Warrior and the stumbling block of the Sovereign'.

At home there were visits from cousins in the Services, one, Captain John Ferguson, instructing at Sandhurst, and another, Flying Officer Jack Gilmour, to be killed in a Lancaster bomber over Europe, followed shortly by his younger brother Hamish.

All this was typical of the average British family at the time. Already I had lost friends from school killed in action and we all lived with stark tragedy affecting friends, relatives and neighbours as part of the common lot. No wonder the country was so unified and determined.

During these short periods of leave I used to have long talks with my father who took an intelligent and friendly interest in

27

all I did. My mother, who was a magnificent supporter and taught me from the earliest age to fight back against disappointment, would leave us sitting together while she fought the interminable battle of the wartime housewife against shortages and rationing—but somehow produced astonishingly good meals without any fuss.

Besides his rigid integrity, my father also possessed great charm and courtesy. He had, like all men, the vices of his virtues and I sometimes mistook his natural modesty for lack of initiative. But he belonged to a magnificent generation, the survivors of the First World War. His friends were not only in the City but on the Continent, where salvage operators and ships' masters held him in high esteem. One German naval officer—the First Officer of the liner *Bremen*—who visited us at Norbury before the war was named Hans Langsdorf. A few years later he was to commit suicide in Montevideo after having had to obey orders to scuttle his ship rather than face renewed action with the Royal Navy. His ship was the battleship *Graf Spee*.

Despite his strong and conventional religious faith, my father was no prig. He enjoyed good company and good conversation and was a considerable *bon viveur*. Through his efforts, our fortunes had prospered, but we were never particularly rich. Even so, it was one of his greatest pleasures to take the whole family up to London for lunch or dinner at Simpson's in the Strand or the Trocadero or to stay at a rather grand hotel at Eastbourne.

So, in his quiet way, he showed me that there was a lot more than English etiquette to the living of a full and successful life.

I had been kept at Maidstone because I was too young to go before an Officer Selection Board, but this procedure was allowed at eighteen. I passed and was off to 164 Infantry OCTU at Barmouth in North Wales. After four months of training I passed out, first of my term, and was awarded the Belt of Honour.

This gave me the chance to fulfil my greatest ambition. Although I had named the Argylls as the regiment I wanted to join, with the Black Watch as second choice and the Gordons as third, there was no guarantee that I would be able to do so. However, the officer cadet who passed out first was given the privilege of choosing his regiment. My most cherished hope would soon be fulfilled.

At the passing-out parade, all but the winner of the Belt of Honour march past as officer-cadets, whereas he was already deemed to be commissioned and entitled to wear the uniform of his regiment and the insignia of Second Lieutenant. This presented innumerable problems for a cadet going into a Highland Regiment, even in war time, and I borrowed bits and pieces from every military tailor who had a representative in Barmouth and a kilt from a friendly Argyll officer on the permanent staff.

On the day of the parade the General who came to present the coveted Sam Browne started off well but put the cross belt over the wrong shoulder. 'By Golly,' I thought, 'what shall I do now?' I cleared my throat apologetically and said 'Sir, I'm afraid you've got it wrong!' He gave me the sort of look I was to see in the eyes of a number of General officers over the next twenty-four years.

From Barmouth I went to the joint Black Watch/Argyll and Sutherland Highlanders Training Centre at Perth. It was commanded at the time by an Argyll officer, Colonel George Mackellar, who saw me off to a marvellous start in the Regiment. He arranged for me to go to Italy and join our Argyllshire Battalion, the 8th Argylls. At this time all officer reinforcements were going to the British Liberation Army in France.

Our training continued through the winter of 1944 in the 10th Battalion Black Watch at Lockerbie. For the first time I had a proper platoon of my own. We trained hard on snow-covered Scottish hills to a peak of physical fitness in readiness for the action I hoped soon to join.

But the war in Europe was ending and all through Christmas of 1944 I champed at the bit as we trained yet again and attended a Battle School. With infantry platoon commanders at a premium and a fifteen per cent casualty rate in Europe at the time, while the allies prepared to cross the Rhine, I stuck out to get into the 8th Battalion which was still fighting with the Eighth Army in Italy. The kind Colonel Mackellar fixed it with the War Office and gave me a note.

This note—it was no more than a single sheet of paper from a scratch-pad—was to become a talisman. It included the words that, if I could reach Italy, 'you should write to Lt. Col Alec Malcolm commanding the 8th and say I have told you to get in touch with him. I will write him now and tell him to apply for you . . . They have recently been near Bologna.' Thus throughout the long series of obstacles that had to be overcome by persuasion and cunning, I was able to say impressively, 'I have a personal message from Colonel Mackellar for Colonel Malcolm, which I must deliver by hand.'

This was the scrap of paper that finally got me to war.

CHAPTER 3

'*There should be a new name for war, because there is no glory in war any more. The word that should take its place is certainly a hard one to find, and must be a word devoid of decency and of sense, carrying no sonorous echoes of tragic beauty and trailing no clouds of glory. I do not know what the word would be, but you must help to find it for war is a filthy and unclean thing. It is only the men who, faced with the necessity of seeing it through, put up with the misery and filth and agony of it—it's only those men who are glorious. The war they have to fight is not—and they'll be the first to tell you so; and if it is ever to happen again, then that fact will always be a shadow shutting out the sun.*'

STANLEY MAXTED in a BBC talk in 1945.

I LAY ON the floor of a railway cattle truck and wondered what the distant strange rumbling noise could be. It was a sound I had never heard before and it was the sound of battle. I was in Italy and now, at last, I knew that I was about to achieve my ambition: to go into action with the Argyll and Sutherland Highlanders.

This ambition had been difficult to achieve. I had thought that the Army would be only too pleased for a keen and highly-trained young subaltern to go to battle. But this did not seem the case. I had had to overcome formidable obstacles to get posted to Italy at all; the war in Europe was nearly over and those officer reinforcements who were not sent across to Germany for the battle of the Rhineland were destined for the Far East. But finally I had set sail in the troopship *Georgic* from Liverpool. At first we sailed in convoy, heavily escorted because in those last months of the war the German U-boats had returned in strength to British home waters and the shipping routes. But, after Gibraltar, our ship had steamed ahead alone and I had landed at Naples.

Here my problems of reaching the fighting began again. First, I was posted to a Reinforcement Depot, a place full of misfits, dodgers and men under suspended sentence—the dregs of the Army, it seemed, and certainly not a unit I had expected to join

31

In despair, I felt that I might be kept waiting there while the war was fought to its finish in the north of Italy. I had the luck to run into a wild young subaltern of the Argylls who was bound for Popski's Private Army, and together we managed to engineer our onward move. Now my 'urgent personal message' to Colonel Malcolm was to prove its worth. First, we boarded a small troopship—it had been Mussolini's private yacht—at Brindisi and steamed up the Adriatic to Ancona, where we were put into cattle trucks. It was then, lying with my kit on the hard, dirty boards, I heard the strange distant rumble from the north that told me that at last I was going to battle.

I arrived just in time. The 8th Army in the east of Italy and the 5th Army in the west were about to break out from their winter lines in the final offensive designed to smash through the German defences in Italy and burst out into the Lombardy Plains. Then, while the 5th Army linked up with the Americans advancing east from France, the 8th could sweep round into Yugoslavia and northward into Austria to meet the Russians. It was a moment of history to stir the blood and my excitement grew intense as the sound of the guns grew louder.

The 8th Argylls were part of the 78th Division, which wore a battleaxe flash as its insignia. Under the command of Major-General R. K. Arbuthnott, late of the Black Watch, it was destined to play a vital part in the offensive which was about to begin. Despite the Allied command of the air and superiority in numbers and weapons on the ground, the enemy defences were immensely strong.

The present line held by the Division lay partly along the banks of the narrow River Senio. Here the opposing armies faced each other at close quarters—sometimes within grenade-throwing distance—under conditions which cannot have been unlike those in the trenches during the First World War. But only after this German line had been broken would the real test come.

The advance northward would have to be up Route 16, a road of great strategic importance leading to and across the wide river Po. But this road was easy to defend because it ran through a narrow neck of land on the east of which lay Lake Comacchio and on the west farmland which had been flooded by the enemy to make it impassable for infantry, let alone armour. The battles would have to be fought up the road and through towns and

32

villages on a front sometimes only about three miles wide. We knew that the Germans were masters of defensive warfare, with their clever use of highly mobile, self-propelled 88 mm. guns and Spandau machine-guns. The coming battle, although probably the last of the war in Italy, was bound to be fierce.

I found the 8th Argylls on 5th April when they were out of the line and training with the 9th Lancers, who were to give them tank support in the coming attack. At last I had achieved my goal and joined the battalion of my father's old Regiment which came from his own hills of Argyll. This was a thrilling moment. The reputation of the 8th Argylls was legendary. They had battled from France in 1940 in the 51st Highland Division; gone to North Africa with 78th Division where they had recaptured Longstop Hill and where Major Jack Anderson was awarded the Regiment's second Victoria Cross of the war; fought through Sicily in 1943 and the whole of the Italian campaign to date. They were a battalion of exceptional quality.

I had a mental picture of them which fitted this reputation and I was not to be disappointed. But the first impression I got was not of their toughness and experience but of the utter friendliness of everyone. I was made to feel completely at home and entirely welcome. After days and weeks of being with strangers it was a wonderful feeling to be back, and 'at home', with the Regiment. It is this feeling of personal identification which makes the infantry regimental system unique. In the 8th Argylls it had been developed to perfection.

But there were one or two surprises. I had expected to find everyone about the same age as the young Jocks in the training battalion in Scotland. But to my astonishment I saw that everyone was much older. My Company Commander looked really elderly—he was, in fact, over forty. My own platoon was the greatest shock. Instead of thirty or more tough, young Jocks I saw just fifteen tired, hardened old men. At least, to a nineteen-year-old subaltern, in freshly-pressed battledress, they looked like this, although most of them were in their early twenties. It was almost continuous action which had made them like this.

But I also knew I was lucky. My platoon sergeant, a tough little man called Dempsey was obviously just the professional I needed to nurse me into my new job. Not only was he to do this in the weeks that followed, but he was to save my life.

33

The company officers mess was in a farmhouse, and it was supper time. Here I met the three other subalterns, who all seemed the right sort of self-reliant young men, but the atmosphere was, once again, far from what I expected. Instead of talking about the coming battle and straining at the leash in anticipation, the officers were all drinking and chatting about past battles and joking about people I had never seen or heard of. I did not drink but they did not hold this against me. There was one other abstainer. Within a few weeks he was to be dead and I was to be introduced to the cheering effects of alcohol by my brother officers.

As the evening wore on I felt more and more relaxed in their company, and as visitors called in from other companies and one or two started to sing I realized that here, at last, I was enjoying what I had read about, that camaraderie which comes on the eve of battle and is part of the heritage of the fighting soldier. One of the songs was so poignant that I ultimately wrote the words down. It was sung to the tune of the German Afrika Korps song *Lili Marlene* and it ran:

We're the D-Day Dodgers out in Italy,
Always drinking vino and always on the spree
Eighth Army shirkers and the Yanks,
We live in Rome and dodge the tanks—
We are the D-Day dodgers,
The boys whom D-Day dodged.

We hope the boys in France will soon be getting leave
After six months service, it's a shame they're not relieved
We can carry on for two or three more years
And nobody need shed any tears—
For we're the D-Day dodgers
Out in Italy.

If you look round the mountains through the mud and rain,
You'll see rows of crosses, some which bear no name.
Heartbreak and toils and suffering gone,
The boys beneath, they linger on—
They were some of the D-Day dodgers
And they're still in Italy.

There were several versions of this song, varying in ribaldry,

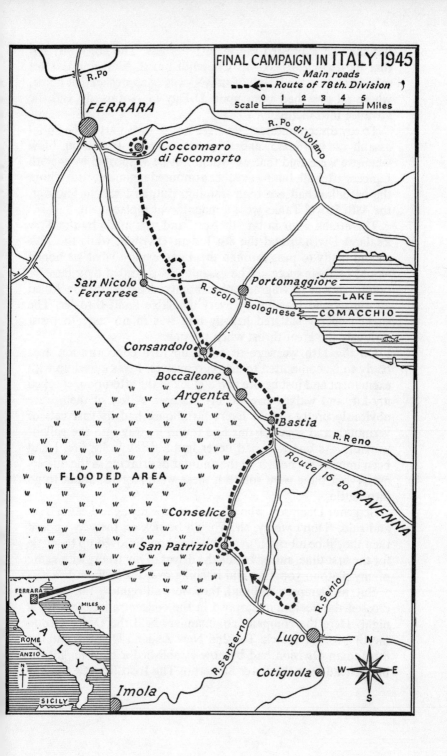

but they all expressed the same bitterness. The 8th Army felt that its thankless task—as Churchill put it, 'dragging the hot rake of war up the length of Italy'—was unappreciated at home, where all eyes had been upon D-Day in Normandy and the advance into Germany.

To my disappointment, we were not to take part in the initial assault on the Senio, and there was another frustrating blow when we were told that we would not be supported by the 9th Lancers after all, but by another armoured regiment, with whom the battalion had not been training. But, as I was to find out, the 48th Royal Tanks were a magnificent replacement.

The attack went in on 9th April and was made by the New Zealand Division and the 8th Indian Division, while the 78th waited ready to pass through them and exploit what we hoped would be their success. The assault was preceded by a massive bombardment of fragmentation bombs dropped by the Allied air forces and blasting by every available flame-thrower. The shaken enemy suffered heavily and was in no state to resist seriously the great thrust when it came.

On the 11th we were to move up to a concentration area ready to be committed to action ourselves. I was keyed up with excitement and just before two o'clock in the afternoon, checked my kit, and walked over to my platoon. A few of them were obviously pretty drunk. Every farmhouse had its vast vats of vermouth and some of this had found its way into the gallant warriors. As one who had never had a drink in his life, I had been impressed enough by the amount of drinking in the officers mess—but these were soldiers, these were Argylls, about to go into battle.

Sergeant Dempsey, who was sober, saw my aghast expression and said, 'Don't worry, sir. They'll be sick on the way up and then they'll be all right.' And this was how I moved up to battle for the first time, sitting beside the driver of my truck, with some of my platoon vomiting and retching over the tailboard.

But, as Dempsey had said, they were all right by the time we crossed the Senio and stopped in the concentration area for a night. Here the company commanders held the Order Group meeting for the attack. The 2nd New Zealand Division and the 8th Indian Division had by now established a hard-fought-for bridgehead over the River Santerno. The Irish Brigade were to

36

go through and the 8th Argylls, who belonged to the neighbouring 36 Brigade, were to protect them from a flank attack, which meant that we ourselves would have an exposed flank. There was talk of the possibility of a German counter-attack with Tiger tanks, which in the event was repulsed by the New Zealanders. However by the time the orders got down to my level we were concerned with moving through the 8th Indian Division late that afternoon to extend the bridgehead, which meant capturing a group of houses simply known as Tre Case a thousand yards beyond their forward defended localities. After this we were to exploit northwards in the direction of San Patrizio and Conselice.

I thought it was terrific and at the 'Order Group' I carefully followed every word. At the end Sandy Alexander, our gallant old company commander said, 'Any questions?' Remembering my Officer Cadet School training, I racked my brains to think of what he had not covered and said 'Where is the Regimental Aid Post going to be?' Everyone laughed—it was obviously the question of a 'New Boy'. I blushed. Then followed more hours of waiting, my first experience of this most trying time for soldiers. The battalion had been 'married up' with a squadron of armour equipped with Churchill tanks, a troop of self-propelled armoured 17 pounder anti-tank guns (called M 10s) a platoon of 4·2 inch mortars, a platoon of medium machine-guns and a troop of Royal Engineers. It was very much the sort of line up that I had trained for at home during the last eighteen months of the Second World War.

At last the order to advance was given and, for the first time, I led Jocks of the Argylls over a start-line. It was one of the most thrilling moments of my life. But exhilaration was soon to be tempered by the first signs of war. At the roadside lay blood-soaked British webbing equipment torn from wounded soldiers only a few hours before. We came upon German graves marked by rifles thrust into the freshly-dug earth and topped with steel helmets and sometimes draped with black rosaries.

Then came the first moment of danger. A mortar bomb dropped out of the sky and burst on the road twenty yards ahead. But there was no sign of the German soldiers until we had reached a second river and doubled across a bridge the Sappers had built for us, now passing the fierce, bearded Indian soldiers who had made the first attack. Once over the river we were on our own and deployed with our tanks and advanced through farmland of growing corn and olive groves. Now I saw my first German, a terrified, white-faced boy with his hands up. Somebody said he had been a sniper.

Slowly we moved forward, each platoon with its tank. I had no idea what was happening beyond my own limited circle of vision. Then and there I reached an early conclusion about the causes and conquest of fear. If a soldier knows what is happening and what is expected of him he is far less frightened than the soldier who is just walking towards unknown dangers. It is difficult for a platoon commander to know exactly what it is all about but obviously I had to radiate a bit of confidence. So, whenever I could, I nodded wisely and said that everything was going splendidly. It seemed to go down quite well.

As evening came on we passed burning buildings, beautiful great farms which, until that day, had lived in peace for perhaps centuries. In the Middle Ages these plains had been the battle-fields of the Medici, and later Napoleon had passed this way, but for generations all had been peace and fertility. The blazing farms and ravaged farmland made a deeply depressing impression on me and stressed the futility of war. I had a curious mixture of emotion as evening fell: excitement, depression, apprehension, exhilaration and the beginnings of fatigue.

There was the thunder and scream of aircraft engines overhead, the explosions of shells ahead and to either side and, in front, the crackle and splutter of small-arms fire. Darkness fell and the olive groves were lit only by the glare of flames and the flash of gunfire. As left flank protection for the 38th (Irish) Brigade the 8th Argylls were ahead of schedule and it had been decided to delay the breakout of the Irish until the following morning. But as the opposition was light and the battalion fresh, we were ordered to push on for San Patrizio which was the first place of any size on the road to Conselice. We went on advancing through the night and drove some of the way sitting on top of the Churchill tanks.

I found it very confusing in the dark to know what was happening but the other two platoon commanders, Alex Lamont and Andy Durward, were terrific and whenever I was in doubt one of them helped me out. Someone captured an 88 mm. self-propelled gun, complete with crew, who had got out to ask some Jocks on a Churchill what it was all about—they had mistaken it for a German tank. The trees at the side of the road had all been prepared for felling but none had fallen so obviously the enemy had been taken by surprise. Out of the garden of a house walked a German soldier. I recognized the shape of his helmet. Quietly he put up his rifle and ordered 'Halt!'

I was frozen with amazement as I was only a few yards behind one of my marching men in platoon headquarters and too stunned with surprise to use my own sub-machine gun. The two of us—German and Scot—just stared at each other. Then, from the darkness behind me, Sergeant Dempsey sprang like a cat and with one wrench had torn the gun from the German's grasp. A moment later, the soldier who might have killed me was being chased back along the street with his hands in the air.

Relief and shock were overwhelemed by shame. I had been tested and I had failed. I had met my first enemy face to face and I had just stood there like a hypnotized rabbit. But for Sergeant Dempsey's speed of action I would be dead. I trudged on in the darkness, sunk in gloom but determined to do better next time.

Our advance had paid off and been too quick for any really organized enemy resistance. But now he began to fight back and we had some exciting moments riding around on the Churchills firing our automatic weapons at enemy positions until they were knocked out or overrun. The total battalion bag of prisoners was twenty and three enemy tanks knocked out as well as the SP gun captured. Later that night we continued scrapping for possession of two bridges to the west of the village and three enemy were killed and a further two prisoners taken. A small Volkswagen staff car was ambushed and an enemy armoured car tried to rescue it. Altogether it was quite a night and by dawn the total of prisoners was 80 for the twelve hours we had been engaged. The following night we extended 2,000 yards to the west and captured 100 prisoners. The 8th Argylls had had 5 men killed and 11 wounded.

The Germans were beginning to fight back with that military professionalism which had sustained them for five long years of war. As always in Italy they depended on the use of self-propelled guns and small groups of infantry with Spandau machine-guns as the basis of their defence. We had that massive air support

39

which was the most effective answer to his stubborn defence. Small parties of Germans, long after their flanks had been by-passed, still carried on fighting. We went down to another position while a fierce German counter-attack was mounted from out of Conselice. Seven more prisoners were taken and it became apparent that the enemy had withdrawn. To the south the New Zealanders were clearing up, but a patrol from 48th Royal Tanks pushing on northwards found a lot of mines in their path. The opposition was still active but on our right the Irish Brigade were now well launched and some reached Reno, south of Bastia. The whole front was loosening up and we were now halted to be rested while the Brigade Area was handed over to 2nd Commando Brigade who were preparing to use a new landing equipment known as Fantail and attack the enemy on the right flank of the Argenta Gap.

That night we dug in round a farm and mostly lay on the floors of the barns and slept. I walked round seeing that all was in order. From a small air raid dug-out in the garden of the farmhouse I heard a strange noise. I climbed down some steps and saw one of my men making love to an Italian woman. I was stupefied with embarrassment. I had never seen anything like this before and all my Presbyterian upbringing reacted in revulsion. I was shocked too, because the man was a most trusted soldier and I knew him to be married. As I lay down on the hard ground to sleep, I resolved to give him a stern lecture on morals. I told Dempsey I was just going to doze off for half an hour. I slept like a dead man. When I awoke I first thought I had slept for thirty-six hours but it was probably nearer twelve—and when Sergeant Dempsey woke me I had other things on my mind than moral strictures.

At the time of the capture of Conselice we were the forward troops of the Army but the front quickly moved on and we had to travel quite a distance in trucks to catch up for the next phase. This was the German key defensive position, the Argenta Gap, on which they hinged all their river defence lines. It covered the mighty river Po and was so named because the town of Argenta was the centre of a narrow strip of land only three miles wide and four miles in depth running between flooded land impassable to tanks or infantry. It was heavily fortified, mined and sited with all the cunning of the German military mind. But it had one weakness. The Germans did not have enough soldiers to man the defences properly and were rushing a new Division down from the North of Italy to remedy this shortage. It was therefore a question of whether the 8th Army could break through before these reserves arrived. By the afternoon of the 15th April 56th and 78th Division attacked and had entered Argenta. But this was only half-way through the defile and so that evening 36th Brigade were committed again. The objectives were the little towns of Boccaleone on Route 16 and Consandolo, defended by Panzer Grenadiers.

We were soon deployed in the Argenta Gap. Sandy Alexander was directing platoon commanders to defensive positions, to dig in and await a possible counter-attack and our own next advance. To me he said, 'Take your chaps into the cemetery and dig in.'

From the purely military point of view, a cemetery, particularly an Italian cemetery, makes a magnificent defensive position. It is surrounded by a massive wall and the tombstones offer good cover for firing positions with shelter from bullets, shell splinters and blast. But to tired and shaken young men, digging slit trenches six feet deep, like the graves of those already dead, this was a grim experience. This cemetery was ghastly and bizarre. There stood the massive white marble mausoleum of the landowner with its locked iron gates, the carved statues of the Virgin Mary and angels, ghostly white in the darkness, the garish crosses decked with framed photographs of the dead and withered lilies.

By dawn, my batman, Private Burt, and I had dug ourselves in. All seemed quiet, so I climbed out of the trench to speak with Dempsey, who was entrenched nearby. But by now the Germans had realized where we were and decided to put down a *stonk* from their *Nebelwerfers*—salvoes of high-explosive bombs, fired from mobile projector batteries, that came in like express trains and burst with shattering explosions. We heard them coming and dived for cover, huddled at the bottom of our slit trenches.

The Germans had the range of the cemetery exactly. The mortar bombs roared down on to us in fives and sixes, the ground rocked around us and the explosions deafened us. Perhaps it lasted three minutes—then it stopped. I lifted my head and looked across at Burt at the other end of our trench. He was standing up and very white. 'I've been hit, sir,' he said. Rocket fragments had caught his shoulder and his shirt was soaked with blood. I helped him out of the trench and back towards the casualty aid post.

After the holocaust there was a strange silence. I walked out of the cemetery gates and on to the road. There stood a troop of light armoured cars of 56 Reconnaissance Regiment, their commander, a young subaltern like myself, leaning out of his turret. He seemed very confident and replied to my questions, 'The Teds are on the run.' The 8th Army always called the Germans 'Teds', from the Italian word 'Tedeschi'. 'We're just going down the road to have a look.' With a grin and a wave, he was off, the cars rolling on down the deserted dusty road towards the enemy.

Half an hour later, a Jeep came back down the road and lying on a stretcher, laid across its bonnet, was a bloody bundle of blanket. Somebody said, 'That's the Recce troop commander. They got brewed up just down the road.'

Now, at last, we were ordered to leave that awful cemetery. With relief, we again moved forward on foot through growing corn, by-passing the town of Argenta and already half-way along the vital isthmus, without having run into any really serious opposition. I was still annoyed with myself over the incident with the German sentry but I soon began to realize that people had forgotten I was a 'new boy' and I was being allowed to take my place as a platoon commander. The confidence this brought stemmed from watching everyone else in authority. Apart from the 'X' Company officers I was immensely impressed by the Battalion Intelligence Officer, Gordon Munro, who seemed to be everywhere, knew exactly what it was all about and was utterly unrattled by the battle. I thought to myself, 'If you can be as relaxed and as cheerful as that chap you will be OK.' I had the usual new boy's dreadful fear of failing and I was much more frightened of that than any of the horror going on around me.

We moved at 5 a.m. on 18th April 1945 with Consandolo as our objective. We soon got involved in a battle the West Kents were having for Boccaleone. Mortar bombs began to thump down amongst us and, from somewhere ahead, enemy machine-guns opened up. Everybody dived for cover and Dempsey called out, 'I think we're pinned down, sir.' I had been trained to believe that the expression 'pinned down' was greatly over-used and, on this occasion, I decided to prove it incorrect by walking over to company headquarters. This was in a ditch about two hundred yards away, across an open field. So I stood up and walked across the field, bullets snapping past my head and kicking up earth at my feet. I reached the ditch without being hit, had a brief word with Sandy Alexander who was busy talking to the CO on the wireless and asking for artillery support, discovered where our other platoons were deployed, and got up to walk back. As I started, one of the Jocks called out, 'You'll be all right this time, sir. They can't see your map case.' Then I realized that I had been carrying this the wrong way round, so showing its shiny side rather than the dull khaki cover. This had caught the sun and the eyes of some Germans some way away. Perhaps I was not all that well trained after all. I grinned rather sheepishly at the company commander and walked back through the corn feeling very lonely and a bit stupid. I joined my platoon on the ground and accepted that we were 'pinned down' as long as the platoon in front of us seemed to have reached the same conclusion. But after about ten minutes of improving our position I decided to go forward to make contact with Alex Lamont's platoon to discuss our next move. They were deployed around a farm and I walked towards two haystacks about fifteen yards apart and perhaps twenty yards in front of me. At that moment a figure darted between the haystacks—but it stopped half-way, stopped dead by a rip of machine-gun fire. There, caught by a burst of Spandau fire and stone dead, lay Alex Lamont, another brave young man to die, after years of campaigning, in the last days of war. Duncan, an officer from the HLI who had joined with me so enthusiastically only a few days before, had also been killed.

This was still the 18th April and now, for the first time, my platoon was to lead the company advance. My fears of failure vanished and, despite the fatigue of a week in action, I was already beginning to feel the confidence that comes with experience.

At noon on that day, we were advancing on Consandolo parallel with Route 16 beside three Churchill tanks in the hot spring sunshine. We were in scraggy countryside with trellised vineyards and to our left we could see the high embankment of the road and ahead the rooftops of Consandolo. Apart for the occasional burst of a shell in the distance or the scream of a diving aircraft and the steady rumble of our tanks, it had become a quiet afternoon, without the noise and tension of the earlier morning.

At that moment the leading Churchill stopped with a violent metallic clank. I assumed it had fired its gun but there was something odd about that tank: it had changed shape. All the great boxes of spare ammunition, the jerricans of water, the bedrolls and the bundles of camouflage netting that hung on its sides had instantly vanished. Then another great clang and the second tank stopped and the same sudden thing happened to the third.

Standing beside the tanks in the lane I then realized that they had not fired their guns, but the Germans had fired theirs and all three tanks were knocked out. At that instant, three white faced men leapt out of the nearest Churchill's turret hatch and with quick jerky movements ran to the ditch where they flung themselves on the ground, lying with mouths open and eyes staring—obviously very shaken. I remember thinking that it all looked rather comic—like a Keystone Cops comedy sequence.

But there was nothing funny about what had happened. A very young Jock beside me in the ditch fell quietly back and lay looking at the sky, a hole no bigger than a sixpence in his forehead. At first I thought there must be a sniper at the other end of the ditch. Then I realized that it must have been a fragment from one of the 88 mm. shells which had hit the tanks. He was quite dead.

Something even more horrible now happened. From the turret hatch of the nearest tank appeared a face, ghastly white and drained. Two hands grasped the edge of the hatch and slowly the man dragged himself clear. He tipped himself over the edge and began to slither down the outside of the tank and would have fallen on his head had not two of those who had just escaped jumped forward and caught him. I saw then that his battledress trousers were a bloody mess: his left leg had been severed at the thigh, his right below the knee. He had been the tank driver and, for the past five minutes, had been dragging himself from his seat low down in the front of the tank to the hatch at the top.

No one came out of the other two Churchills and we could only assume that they had been killed by the solid, high-velocity shells shattering the inside of the armoured hulls.

The legless tank driver asked when the stretcher bearers would be arriving and I said soon, but he gave a bitter little laugh of disbelief. I felt I must cover the face of my dead Jock and the terrible wounds of the driver. I went across to one of the stopped tanks and picked up a bundle of greatcoats that lay on the ground—but they came apart in my hands like dust, disintegrated by blast. Finally I found some tattered pieces of blanket. One I put over the dead boy's face the other over the tank driver's legs, but he pushed it aside, seeming fascinated by the horror of the sight. All he did was to ask me to brush away the flies which clustered on him.

But we had to move and I could only leave a Jock to comfort him and shield him from the exploding ammunition in the tanks which were now beginning to burn. Later, I heard that he had died.

The battle that raged all that day around the villages of Boccaleone and Consandolo destroyed the enemy in the Argenta Gap and the system of deep defensive lines upon which he had placed so much store to defend the river Po. We were now fighting the 26th and 29th Panzer Divisions. It was ferocious fighting and both sides suffered heavy casualties. In the 8th Argylls alone the casualties amounted to fifty that day. It was war as I had somehow expected to find it except that it was so much more concerned with small, somewhat pathetic groups of people. I had thought I would be able to influence events and instead I was dominated by them. It was ghastly, except for the courage and comradeship of the people around me. I remember thinking again that war was utterly futile and thank goodness this one was about to end.

The enemy were now in full retreat although some rearguard elements made vicious local counter-attacks. The bag of prisoners mounted and the battle on the 8th Army front was beginning to develop into a chase. The 6th Armoured Division had swept forward along Route 16 and 78th Division were to head due north. By 19th April it was reckoned that the Germans were holding the Scolo di Porto river obstacle with all the strength they could muster. On the 20th we were told they were

45

withdrawing again and it looked as though the Po was the next objective. The 8th Argylls were ordered to seize a bridge over the Po di Volano, three miles east of Ferrara. At 3.30 on the afternoon of 21st April we were off again. The country was very flat, interspersed with numerous small canals and rivers but much more open than the olive groves and terraces of the previous days. It was good tank country and the Armoured Corps made the most of it; we moved forward very rapidly, but not without cost. The Argylls suffered another twenty-five casualties.

I was wounded myself but not included in that total. We were crossing an open field, when, only about three hundred yards ahead, I saw a German self-propelled gun drive out from behind a haystack and dash off behind a line of thin trees. Almost instantaneously I felt a violent blow on my hip, put my hand down and felt hot blood. I told Dempsey I was all right because the fragment of shrapnel had hit my mess tin which held a very solid block of condensed chocolate—our 'iron rations'—and this had taken some of the impact. But later I was persuaded by Dempsey and the others to walk back to the aid post.

The Regimental Aid Post was in a barn and I entered to see a dreadful sight. The great barn was crowded with wounded: infantrymen, burned tank crews, Germans, some lying with fearful wounds, others standing or leaning against the walls, while doctors and medical orderlies did their best to help them. With so many in far worse condition than myself, I could not stay. I felt ashamed even to suggest that I needed help. So I limped outside and put a field dressing on my wound, leaving the shrapnel in the flesh, where it has remained ever since.

There is an ironical postscript to this incident. After the war I was informed that, although I was entitled to the Italy Star campaign medal, I was not entitled to the 1939–45 Star. The qualifications for this were to have served in a theatre of war for six months or be decorated or wounded in action. I said that I had been wounded in action but was told that my wound had not been entered in my records and I was therefore ineligible.

We went on through the evening with occasional mortaring and as darkness came we continued to press on. The tank crews struck me as being utterly fearless in the way they ignored the threat of an unseen enemy armed with *panzerfaust* and mines. Just after midnight we reached the company objective and seized our bridge intact. We had moved seven miles, most of it in the dark, a classic example of infantry–tank co-operation. For the rest of the night we cleared up various enemy who were escaping or surprised by our arrival. The next morning I put a section into the upper attic of a large farm covering the river and on visiting them an hour later was annoyed to find them less attentive than I thought the situation merited. I showed them how I wanted things done and as I finished, looking between slits in the woodwork I spotted a German soldier about fifty yards away apparently standing beside a slit trench. Taking the rifle from the nearest Jock I carefully lined him up in my sights and fired. He did absolutely nothing, seemingly unperturbed. I reloaded the bolt action of the rifle and as he sauntered away I fired again. Still unhit he disappeared from view. The Jock looked at me with a quizzical grin. 'Plunging fire,' I said, 'make sure you do better, it's your bloody rifle anyway!'

The next day we were off again to advance on Ferrara from the east. To the 5th and 8th Army this famous old Italian city had been a coveted prize ever since the fall of Bologna. To me this climax of the battle for northern Italy was a scene of extreme danger and farce. Two companies of the 8th Argylls, while moving up towards the Diversio del Volano, had assembled in a large farmyard, while the company commanders discussed our next moves. As perhaps two hundred of us went about our various business, a surprise visitor arrived. Out of the olive groves and into the farmyard burst an enormous German tank. We all ran like hares. Company headquarters sprang into a deep ditch filled with stagnant water, so that only their heads showed. Others, myself included, began to run round behind a huge barn with the tank, waving its long gun, rumbling in pursuit—and again, with odd detachment, I thought of the Keystone Cops.

We ran right round that barn with the tank after us, hared across the open farmyard and flung ourselves to the ground. One officer, Ian Gunn, and a Jock had meanwhile taken action. They had got hold of a small Piat anti-tank bomb-projector and were lying in the open, training it on the corner of the barn, where the German tank was about to appear. When it did Ian squeezed the trigger. The bomb just popped off the projector and dropped a foot or two in front, whereupon the two brave Argylls took

47

to their heels and leapt into the stagnant ditch with the others.

Again I found myself mesmerized by the strangeness of it all. The huge tank swung into the big yard and, as it did so, a German, wearing a rather incongruous steel helmet, stood up in the turret hatch holding a stick-grenade ready to throw. At this moment, Sergeant Dempsey did what I should have done. He suddenly broke cover, raced up behind the tank, leapt on to its back, shot the German with his carbine and tried to push a grenade down the hatch. It was all over in a second or two.

The grenade either did not explode or the Germans' corpse had stopped it doing its work, because the tank turned and lurched away into the olive groves, its commander thoroughly scared. Later it was found abandoned and we heard that it had, in fact, run out of heavy ammunition and had been as surprised as we were at our meeting. But we were not to know this. Sergeant Dempsey was awarded an immediate Military Medal— the last the battalion was to win during the Second World War. It was an object lesson to me, showing that courage and quick-thinking were not automatic reactions.

Ferrara was taken, albeit by the troops of 8th Indian Division and Colonel Freddie Graham's 1st Battalion of the Argyll and Sutherland Highlanders, and our combined pipes and drums beat retreat in the main square before crowds of Italians, jubilant at the ending of the war and amazed at the sights and sounds. We prepared for the crossing of the Po but meanwhile 6th Armoured and 2nd New Zealand Divisions had crossed against relatively light opposition and further west the American 5th Army had breached the river line obstacle. The war in Europe was almost over.

The German defences were now crumbling everywhere and peace negotiations, of which we were unaware, finally resulted in the announcement of the end to hostilities on 2nd May.

The 78th Division went mad with delight. That night we fired off tracer, pyrotechnics and almost all the less lethal ammunition we had with us, illuminating the sky as few battles had done. That night I had my first drink—sweet vermouth, no less.

I had been at war for only a month but I was a different person as a result of it. I had aged and matured drastically. But what I had seen and experienced haunted me and, a year later,

sitting one night on guard in Jerusalem under the walls of the
Old City, I jotted down a piece of verse to exorcise my feelings.
It was very much the writing of youth, but the first verse honestly
described how I felt at that time and at the age of nineteen.

'You before were young and fresh and knew so little
Now your eyes have seen too much
You before thought war and death exciting
Yet your senses now are hard to touch.'

CHAPTER 4

*'So I returned and considered all the oppressions that are done
under the sun: and behold the tears of such as were oppressed, and
they had no comforter; and on the side of their oppressors there was
power; but they had no comforter.'*

BOOK OF ECCLESIASTES

I WAS WALKING through the streets of Jerusalem with my
friend John Penman, my company commander in the 1st
Argylls, on our way to lunch. It was a hot July day. About once
a fortnight we treated ourselves to a splendid meal in the grill
room of the sumptuous King David Hotel which was now our
destination. As we strolled along the street the only sign that
1946 was a year of trouble in Palestine was that we wore pistols
at our belts.

The King David Hotel, where we would be ordering from the
elaborate menu in ten minutes' time, was one of the great hotels
of the Middle East. But there was something unusual about it.
When you entered the spacious foyer you could turn right, as
we planned to do, for a drink at the bar, or go down to the grill
room in the basement. But, if you turned left you came upon a
Military Policeman and a guarded door, behind which lay the
British General Headquarters, Palestine, which occupied a wing
of the hotel. As a junior infantry captain I had never had occasion to turn anything but right.

We were within three hundred yards of the hotel when it blew
up. With the most shattering explosion I have ever heard, the
King David Hotel vanished in a colossal billow of smoke and
dust. John and I exchanged glances, knew that we must get
back to the Battalion, turned and ran through the streets now
filling with running troops and police, racing fire engines,
ambulances and police cars: a scene of utter turmoil.

We had just seen—and narrowly missed death in—an ulti-
mate act of terrorism. That morning, Jewish terrorists had rolled
milk churns filled with high explosive into the basement kitchens

50

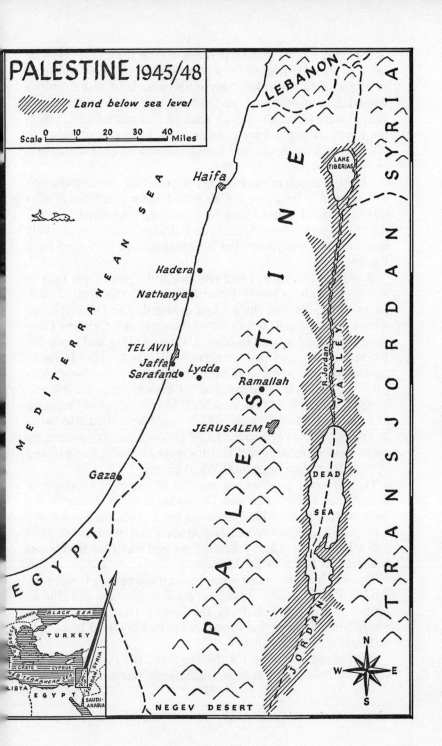

PALESTINE 1945/48

Land below sea level

Scale 0 10 20 30 40 Miles

LEBANON

SYRIA

LAKE
TIBERIAS

Haifa

MEDITERRANEAN SEA

Hadera

Nathanya

TEL AVIV
Jaffa
Sarafand   Lydda

Ramallah

R. Jordan

JORDAN VALLEY

JERUSALEM

DEAD
SEA

Gaza

EGYPT

PALESTINE

TRANSJORDAN

RUMANIA   BLACK SEA

GREECE

TURKEY

CRETE   CYPRUS

MEDITERRANEAN SEA

LIBYA   EGYPT   SAUDI
ARABIA

Jordan
Valley

N
W   E
S

NEGEV DESERT

of the hotel, adjoining the grill room, and set the fuses to explode the charges at lunchtime. The plan worked perfectly and now, in and around the rubble, lay ninety-one dead and forty-five wounded. What was so particularly horrible was that these victims were not solely British staff officers and soldiers, whom the terrorists might have considered as justifiable targets, but also fellow Jews, Arabs and foreigners with no connection with events in Palestine.

The blowing-up of the King David Hotel was my introduction to terrorism at its worst and an introduction to a form of warfare in which I was to become increasingly involved over the next twenty-one years. My stay in Palestine taught me not only how to counter terrorism but how it should be prevented from starting.

A year before, when I had returned to England from Italy as a young subaltern newly blooded in battle, this sort of campaigning was the last thing I had expected. The 1st Argylls, to whom I was posted, were part of the great 6th Airborne Division, which had landed in Normandy on D-Day and made the Rhine crossing, and had been assembling in England in readiness to take part in the re-capture of Malaya and south-east Asia from the Japanese. The division, which was up to its full war strength of about 18,000, consisted of two parachute brigades and one air landing brigade, which would be carried into battle in gliders. The Argylls were to be glider-borne troops and we spent some interesting and exciting weeks learning to fly in and operate from American-built Waco gliders.

Despite great pressure we resisted all attempts to make us discard our Glengarries and Tam-o-Shanters in favour of the famous 'Red Beret' which was worn by all Airborne Forces. We were Argyll and Sutherland Highlanders and much as we liked and admired the 'Cherry Berries' we had our own traditional headgear. It had a longer history.

The war had just ended when I joined the battalion but, with the Far East in chaos, we still expected to go there and help to restore order, particularly in the former Dutch East Indies, where British troops had found themselves involved in an ugly little war with nationalists.

But instead of sailing for the Far East, the division was diverted to Palestine, where an increasingly dangerous situation was brewing up, a situation directly linked with the war in which we had been fighting a few months before.

The struggle between Jew and Arab for the possession of Palestine had been slowly and steadily mounting since the British Government announced in the Balfour Declaration of 1917 that it favoured the setting up of a Jewish national home in that country. Palestine itself was to remain a mandate to Britain which was therefore responsible for the maintenance of law and order, an increasingly difficult task as pressures built up on the opposing sides.

Trouble could be held in check whilst the vast majority of Jews remained settled and assimilated in the countries where they had eventually settled after their expulsion from Palestine by the Romans in the second century A.D. But Nazi anti-Semitism and the massacre of six million Jews in Europe—about seventy-five per cent of the total Jewish population in German-occupied countries—changed all this. There was an intense desire among the survivors of these horrors to escape from the scenes of their suffering and build up a new and independent life in a Jewish state—and this, of course, they saw as being their ancient homeland.

In June 1945, the Jewish Agency, which was the Jewish 'shadow government' in Palestine, sent a memorandum to the British Government demanding the immediate issue of 100,000 immigration certificates to European Jews wanting to settle in that country. This proposal was violently opposed by the Arabs. The United States, still gleefully assisting in the erosion of the British Empire, made matters worse by backing both sides. Washington supported the Jews because of the size and voting influence of American Jewry, and the Arabs because of recent discoveries of oil in the Middle East.

The British Government was particularly anxious to keep the peace. There was even talk of transferring the main British strategic base in the Middle East from Egypt to Palestine. Fortunately, common sense prevailed for this would have been a gross error of judgement, like similar misjudgements to take place over Cyprus, Kenya and Aden. It is not possible to secure an effective military base without complete local political co-operation.

There was now deadlock over the issue of Jewish immigration, with the British Government trying to arrange a compromise

53

between the Jewish Agency and the Arab League but with rapidly fading chances of success. With the increasing tension throughout the summer of 1945, it was decided to send a major British Army formation to Palestine on a peace-keeping mission. In October and November, the 6th Airborne Division landed at Haifa.

Our own feelings in the Argylls were as confused as the political situation itself. On the one hand, we had the greatest sympathy for the Jews because of their recent sufferings in Europe, and were anxious to help them towards a new future. On the other, we had absurdly romantic ideas about the Arabs—passed on to us by my boyhood hero Lawrence of Arabia and the desert explorers of the nineteenth century—and the illusion that because we liked the Arabs we thought the Arabs liked us.

What we saw as a Jewish rebellion—aimed specifically against the limitation of Jewish immigration by the British—had, in fact, begun some years before and, although the first trouble had been on a small scale, it had seen the organization of the formidable Jewish terrorist and defence forces we were now to encounter. I was particularly interested in the influence on their policy made by the late General Orde Wingate, whom I considered the most interesting British commander of the Second World War.

The main force, to which most men of military age belonged, was the Haganah, which itself contained the Palmach, a striking force or *corps d'elite* from which the leaders were drawn. Two extremist groups had broken away from the Haganah in 1940 to form ruthless terrorist organizations which, they were convinced, could achieve more by acts of extreme violence and savagery. These were the Irgun Zvai Leumi (IZL) and the even more extreme Stern Gang. Although separate organizations, their commands overlapped and they would often act in liaison.

We witnessed this when we arrived in October 1945. The three forces acted in concert to stage an act of combined protest against the British immigration policy. The Palmach sank three small naval vessels in Haifa harbour, the IZL sabotaged Haifa oil refinery, and the Stern Gang blew up Lydda railway station. Tension mounted and two months later the IZL attacked several Palestine Police stations, causing nine British casualties.

The Argylls, at war strength of 52 officers and well over

54

1,000 men, were stationed at Camp 21, about thirty-five miles from Haifa and only five from the attractive coastal resort of Nathanya. Under the command of Lieutenant-Colonel R. H. L. 'Squire' Webb, a brave officer of the old school, we settled down to a curious routine in which the new skills of Internal Security operations were combined with both the tough, rather slap-happy wartime attitude to soldiering and the beginnings of the restoration of pre-war peacetime Regimental life. More time was spent on training and sport than on operations during our months at Camp 21. But at Hadera, a nearby Jewish settlement, we had our first experience of Jewish determination when we surrounded the area in order to search for weapons and were thwarted by the effectiveness of their wired-in houses and barns.

We were new to this sort of situation. These people were honest, hardworking farmers and labourers and we could not use force against them. All they did was to stand silently and glower at us from behind locked gates which had been constructed for a siege. We looked out of place sitting around the settlement with our huge armoured self-propelled guns in support. As a Platoon Commander I was glad that the decision as to what to do next was not mine. 'Squire' solved it in his inimitable way by ordering the orderly bugler to blow the Regimental charge. This soldier, wearing a kilt and an airborne steel helmet, blew for all he was worth, and we advanced towards the central gate. The Jews decided to open it. Our mission completed, the officers lunched from white-clothed tables while the mess truck dispersed refreshments under the expert eye of Sandy Bardwell, the second-in-command. It was very much the pre-war form and tradition.

'Squire' Webb was a great game shot and under the expert tuition of Hector Macneal, whom I had known in Italy with the 8th Battalion, I was introduced to the mysteries of ornithology and shotgun shooting. My enthusiasm was greater than my skill and it is tribute to the strong nerves of my tutors that they stuck it out despite my dangerous acts. Once, shooting quail in a field of standing crops, I fired at a retreating bird going towards a mound of sand. To the astonishment of us all the 'mound' leapt into the air with a great grunting—it was a camel, resting with its load of sand from the nearby beach. However, it was entered in the 'various' column of my Game Book where it

55

remains today to remind me of many happy hours of sport and relaxation in that beautiful Holy Land.

In March 1946 the Argylls were ordered to Jerusalem, where we were to be based on the Hospice de Notre Dame de France, opposite the New Gate of the Old City, and the Syrian Orphanage, in the west of the city. Here we were at the vortex of the mounting crisis. But the situation was deteriorating everywhere. In April, the Stern Gang brutally massacred seven of eight men of the Parachute Regiment guarding a car park in Tel Aviv. Illegal immigrants were pouring into Palestine by the thousand and being hidden in Jewish towns and settlements.

We now realized that the men we were up against were not seedy little terrorists such as we were to meet years later in Aden. These were hard men experienced in the pitiless and sophisticated techniques of clandestine warfare learned under German occupation in Europe. They were zealots, combining the toughness of the East European with the intelligence of the Jews, and the cause for which they fought was, as they saw it, that of the survival of their race in a nation of their own. In their patriotic fervour, forged in the fires of Europe, they now transferred the hatred they had felt for the Germans to the British, who were thwarting their ambitions and, they believed, abetting their rivals for Palestine, the Arabs. They were a formidable enemy.

Really serious trouble began on 16th June, when Jewish terrorists blew up nine bridges across the Jordan, including four railway bridges, and the Allenby Bridge, and kidnapped five British officers. In an attempt to discover how deeply the Jewish Agency was implicated, the British decided that all its members were to be arrested and their offices and houses searched. This was to be called Operation 'Agatha' and was prepared with the utmost secrecy by the Palestine Police supported by the Army. My own task was to arrest Moshe Shertok, who, as Moshe Sharett, later became his country's spokesman at the United Nations.

Silently, my Jocks surrounded the stone block of flats where our quarry lived and, having identified his windows on the second floor, two of my men and I shinned up the wall on to his balcony like cat burglars. The windows were sealed by heavy steel shutters. so there was nothing for it but to enter by the

WHITGIFT SCHOOL O.T.C.,
1941, AGED 15

LT. COLIN MITCHELL, M.C.,
ARGYLL AND SUTHERLAND
HIGHLANDERS
(THE AUTHOR'S FATHER),
SEATED, FRANCE, 1917

Photo: Imperial War Museum

BRITISH TROOPS CROSSING THE SANTERNO, ITALY, 1945

PALESTINE, 1946: HECTOR MACNEAL, GORDON MUNRO AND THE AUTHOR
AT HADERA

Photo: Imperial War Museum

BLOWING UP OF THE KING DAVID HOTEL, JERUSALEM, 1946

IE ARGYLLS, WITH MINE-DETECTORS, PICKS AND SHOVELS, SEARCHING
IE VILLAGE OF GIVATH-SHAUL AFTER TERRORISTS HAD MINED A ROAD
OUTSIDE JERUSALEM

Photo: Planet

LIEUT. COL. CLUNY MACPHERSON AND THE AUTHOR, JERUSALEM, 1947

MAJOR JOHN PENMAN AT
SARAFAND, PALESTINE, 1947

KOREA, 1950
THE ARGYLLS ADVANCING THROUGH SARIWON

ARGYLLS TAKE COVER ENTERING CHONGJU

B COMPANY 1ST ARGYLLS DURING THE WINTER RETREAT, KOREA,
NOVEMBER 1950

OFFICERS OF THE 1ST ARGYLLS TOWARDS THE END OF THEIR SERVICE
KOREA, APRIL 1951. AUTHOR AT RIGHT END OF MIDDLE ROW, HOLDIN
CROMACH

HE AUTHOR'S WEDDING TO SUSAN PHILLIPS, MARCH 1956: WITH THE
ARRANT OFFICERS, RSM PADDY BOYDE, MBE, DCM, RSM TOM COLLET, MM,
AND COLOUR SERGT. EDDY EDMONDSON

CYPRUS 1958. DISCUSSION AT APHRODITE'S POOL
(LEFT TO RIGHT) SGT. ADAM, CSM 'GINGER' LYON,
THE AUTHOR AND LT. DAVID FLETCHER

A GROUP OF BRITISH OFFICERS IN THE KING'S AFRICAN RIFLES, NANYUK
1961. MOUNT KENYA IN THE BACKGROUND

AUTHOR AS BRIGADE MAJOR,
KING'S AFRICAN RIFLES, 1962

front door in a more formal manner. I rushed up the stairs, brandishing the Luger pistol I had brought from Austria, and hammered on the door, shouting for it to be opened. It was a hectic moment and I half expected a burst of machine-gun fire from within the flat. The door was opened by two gigantic men, who most courteously told us that Mr. Shertok was not at home and we were welcome to see for ourselves. So we raced into the flat, our hosts helpfully opening doors and showing us round. Obviously, he was not here and we left, feeling rather foolish. The prize catch, David Ben Gurion, was also away—in Paris— so, as far as we could see, Operation 'Agatha' achieved little more than further inflaming Jewish opinion against the British.

This had been at the end of June and we had to wait until 22nd July for the terrorists' answer: the blowing up of the King David Hotel.

Our other life continued and 'Squire' and the pre-war regular officers who had arrived taught us the mysteries of the Regiment. This was perhaps frustrating to the large number of officers who held temporary commissions from the war and who eagerly awaited release to civilian life. To be wakened at dawn and ordered to the roof of the Syrian Orphanage where we were all instructed in Highland Dancing by the Pipe Major was bad enough for some of them, but to be threatened with seven days extra Orderly Officer for referring to the Regimental Colours as 'those flags' added insult to injury.

The Argylls had now, for me, become a second family, and I knew that my life lay with them. I had therefore applied for a permanent commission and was sent in June 1946 to the Regular Officer Selection Board in Egypt. When I got back to Palestine 'Squire' was delighted to find that I had been graded 'Outstanding'; his encouragement and support made me appreciate what a great family team the Argylls were

It was the policy of both the IZL and the Stern Gang to take an eye for an eye. The laws which the British judiciary enforced in Palestine had been designed to hold down primitive colonies and not sophisticated Europeans, and amongst the penalties imposed upon those convicted of helping the terrorists was flogging. This the terrorists effectively stopped by kidnapping and flogging a British Brigade Major of the Airborne Division. Later, when the first death sentence on a terrorist, Dov Grüner, was confirmed, a British judge was kidnapped and held as hostage. The High Commissioner, Sir Alan Cunningham, thereupon threatened that, unless the judge was freed within two days, Tel Aviv would be placed under military occupation and martial law, but conceded a stay of execution while an appeal was sent to London. Whereupon the hostage was released. But the appeal was refused and Grüner eventually hanged. In immediate reprisal, the terrorists kidnapped and hanged two British sergeants, leaving a booby-trap bomb on one of the bodies.

Although most of us had arrived in Palestine with sympathy for the Jewish people, their hatred of us and continued actions of terrorism had made us anti-Jewish—but not, I think, anti-Semitic in the broadest sense. With time on my hands to read, I found the Jews fascinating. One of our most intelligent officers, senior to me, was openly pro-Jewish and saw their struggle against us and their future struggle against the Arabs as a splendid example of courage and resolution and the refusal of the human spirit to accept defeat.

On one occasion, towards the end of the mandate, I was being driven by this officer in his Jeep when we were sniped at from an orange grove. My instant reaction was to stop and stalk the sniper, but my brother officer drove on. I angrily asked him why he had not stopped and he replied, 'He was only doing his duty fighting for his country—and you would have killed him.'

I did not share his sentiments, for the horror at the King David Hotel had been followed by an attack of devilish ingenuity on the Argylls. The battalion was constantly engaged in manning road blocks and observation posts and patrolling and, in October 1946, we had set up a road block and positioned a number of Jocks in the centre of Jerusalem to enforce a curfew. It was the Sabbath and because this was a very orthodox quarter of the city and all work was forbidden on this day, the metal blinds over the shop windows were opened automatically by electric time-clocks. Knowing where the Jocks would be positioned, sleeping on a certain stretch of pavement, the terrorists had planted plastic explosive inside the rolled-up blinds of the shops at that point and set the time-clocks to unroll the blinds at 8 o'clock that night.

All seemed quiet in the street, deserted except for Argylls, many of them sleeping on the pavement, when, silently, the blinds began to unroll. Suddenly there was a series of shattering explosions. Nobody knew what was happening. Were we being mortared? Nobody knew how to react. All we knew was that there on the pavement lay a dead Argyll and a wounded officer, a corporal, who was to die from his wounds, and six other wounded Jocks.

Events could move so quickly at that time, that this lesson in terrorism was followed immediately by another lesson in complete contrast, a lesson in public relations.

Now we were to see a demonstration of the power of the Press. The incident at the roadblock had thrown the whole of Jerusalem into turmoil, with all our companies rushing out to cordon and search suspect areas. The Commanding Officer was at Battalion Headquarters in the Hospice de Notre Dame, coordinating the activities of the various company commanders. Meanwhile the centre of Jerusalem was filling with correspondents of the world's Press, including several distinguished journalists. They were getting in the way of our operations and my friend Hector Macneal who was Intelligence Officer, reported this to Battalion Headquarters. The CO ordered the correspondents to be rounded up and brought before him.

A few minutes later the startled correspondents found themselves being rounded up by Jocks under the direction of the Regimental Provost Sergeant, bundled into a truck and driven under guard to the Hospice. There they were confronted by the CO. I gather he told them in no uncertain terms that he would not tolerate their getting in the way of his soldiers and that this must never happen again. The Press were thereupon dismissed.

It was a very understandable reaction by a Commanding Officer trying to conduct a military operation but it set the wires humming and the Jewish correspondents particularly made much of what they considered a high-handed action. We had to put up with a good deal of publicity, knowing little or nothing of the chain-of-command under which the CO himself served nor of the complexities of his dealings with the numerous staff-officers and intermediate commanders who constitute the Hydra of the military machine. As a result of the COs difficulties I believe that the morale of the battalion suffered at a moment when high morale was most needed. I would blame the higher command for this lack of understanding but on the other hand I only had a

59

'worm's eye view' of the circumstances. Whatever the rights and wrongs of this particular case, the importance of the Commanding Officer as the focal point for his battalion was made very clear to me then and was to influence my own actions when, twenty years later, I myself commanded the Regiment in Aden.

Soon after this I went home to attend the company commander's course at Warminster. My father's reaction to my choice of career was that he hoped the peacetime Army had changed. In both world wars, the British Army had been officered largely by men such as himself. In peacetime, it was his view that it tended to become a social profession, suitable for gentlemen with private means and sporting instincts. It was characteristically cautious of him. My own view was that he was old fashioned and that the post-war British Army would be very different from its pre-war counterpart. Most importantly, it offered a life of travel, adventure and excitement in the old Colonial Empire.

While I was on my course from December 1946 to January 1947 the Battalion moved out of Jerusalem but shortly after I returned we went back. I took to our work with renewed enthusiasm and confidence. It was thrilling to operate in a city of such historical significance. It was extraordinary to hear oneself say, for example, 'Sergeant Major, take your men round by Mount Zion and I'll meet you at the Pool of Siloam—I've got to meet the CO at the Damascus Gate.' We were constantly engaged in 'soft shoe' silent night patrols, raids and searches. One of the most bizarre of these was to search an Armenian church for hidden arms while a service was in progress. As the Jocks with loaded weapons crept into the dim church, the monks continued their chanting, taking no notice of us. As we tiptoed through the incense-scented gloom, searching in dark corners and feeling the walls for hiding places, we knew that they were watching us from the corners of their eyes and I wondered whether any of them were disguised terrorists who would suddenly draw guns and the solemn chanting would be interrupted by a gun battle among the pillars.

It was a strange time. We lived among the enemy in their city and, while we took danger for granted, we were ever alert. We would sometimes eat in Jewish restaurants but always with a loaded pistol on the seat beside us. If a waiter dropped a tray, the guns would be in our hands, fingers on triggers.

One day I went into a well-known tailor's shop to choose material for a new suit. Suddenly I sensed that something sinister was happening. One by one, the Jewish customers left the shop and no more entered. I noticed, too, that the tailors in the cutting-room at the back had stopped talking. Were terrorist gunmen coming through the back of the shop to shoot me? I told the salesman that I had to go, but would be back another day. He tried to persuade me not to leave, but by now all my senses warned of imminent danger and I was out of that shop in a flash. If we had searched it we would have found nothing. At this time, a British officer browsing in a bookshop, which I frequented, was shot dead, so my retreat was prudent. Shortly after we always had to move in threes.

The terrorists did not make all the running. For three months at the beginning of 1947 they were to experience counterterrorism at its most efficient. In March of that year there arrived in Palestine a young and much decorated former major in the wartime Special Air Service called Roy Farran, seconded to the Palestine Police. It is not for me to discuss his operational techniques but occasionally we got involved in his secret operations when it was important to provide a diversion. I found this highly stimulating and, getting a bit careless one day when tension was running high, I drove down what turned out to be a cul-de-sac on the outer edge of a Jewish quarter. Suddenly the air was filled with flying stones and I reversed my open jeep, dodging the boulders. Groups of Jews including one zealot on the rooftops were hurling them from all directions. It was a graphic illustration of the Biblical phrase 'they stoned him to death' and I could see why in Palestine, with small bits of rock lying everywhere, this was such a natural weapon for the mob. Severely shaken, I got back to the main road, unhurt except for a few bruises but with a number of large dents in the jeep.

Roy Farran's actions terrified the terrorists. But eventually they began to understand what was happening, identified Roy and were determined to put an end to his activities. To do this by force would have been difficult, if not impossible, because the Farran team were as tough and experienced as any Jewish terrorists. They therefore snatched eagerly at what seemed like a scrap of evidence. A young Jew named Alexander Rubovitz was kidnapped in Jerusalem and disappeared. But there had been a scuffle and at the scene was found a hat inside which were the initials R.F. The Jews demanded that he be charged with murder and Roy, forewarned, escaped over the border into Jordan. Eventually he was persuaded to return and face trial and I was appointed one of his three gaolers in the elaborate arrangements for guarding him which were known as 'Operation Buffer'.

As a gaoler, first having been relieved of my own pistol, I would be locked into Roy's cell with him for twenty-four hours at a time, sometimes being allowed out to play badminton with him under guard. He spent much of his time writing his memoirs, *Winged Dagger*, As I listened to this hard, cold-eyed young man I began to realize that my boyhood ideas about soldiering in the Middle East—romantic Rupert Brooke ideas about writing poetry on the eve of battle and dying like a gentleman—were a thing of the past. We had left the era of Bulldog Drummond and entered that of James Bond.

Palestine was now moving faster towards total chaos. The British mandate was inevitably coming to an end, illegal Jewish immigrants were pouring into the country, and both Jews and Arabs were arming for the battle for Palestine which would surely follow our withdrawal. This battle was, in fact, already beginning. In September, 1947, the Battalion left Jerusalem and went to Sarafand and later Lydda, where our task was eventually to keep Jews and Arabs from each others throats as well as protecting ourselves and the military installations.

The Jews were now fortifying their country settlements and one of our duties was to escort convoys of food and non-military supplies to these settlements through Arab territory. The Arabs accused us of being pro-Jewish, and the Jews when they did not get the protection for which they had asked, accused us of being pro-Arab. At times we were against both—emotionally and in action.

It was difficult to give whole-hearted help to the Jews after their terrorists had blown up a British troop train, in February 1948, with appalling loss of life. As a result of such incidents, it was rumoured that some British officers withheld escorts from Jewish convoys which were thereupon ambushed by the Arabs. The Jews realized that all British protection would end with the mandate in May of that year and were making desperate preparations for war, establishing their own roadblocks and making armoured cars by covering three-ton trucks with steel plating.

The Arabs, too, were preparing. Near Sarafand, where we were based, the British commander of the Arab gendarmerie, an old friend of mine in the Gordon Highlanders called Dick Gammon, one day woke up to find that his entire command had deserted with their weapons. He was heartbroken as he had a great liking for Arabs. On a later occasion, while we were escorting Jewish electrical engineers repairing telephone lines, Arab guerrillas opened fire on us, and were preparing to attack in force, so that we had to reply with Piat anti-tank bombs, causing them heavy casualties.

It was in this short but exciting action that I got an insight into the character of Lieutenant Colonel Cluny Macpherson, who had taken over command of the Battalion. He was an outstandingly buoyant and charming man whose obvious military brilliance had been arrested by five years as a German prisoner-of-war. In the year that I had known him in Jerusalem and at Sarafand I had come to worship his leadership qualities and recognized that these were very special indeed. This day when we were all under heavy fire from the Arab positions he insisted on walking away from the comparative safety of his armoured scout car to perambulate up and down the road with me and to carry on a rather trivial conversation. I suddenly realized he was doing it just to see how good I was! I over-compensated for this by suggesting airily that we both walked up the hillside to try and stop the shooting. He laughed and said, 'You'll get us both killed, but that's how the old 93rd would have done it.' This splendid demonstration of coolness under fire remained in my memory as an example of what a Commanding Officer should be. I was a young company commander and had just taken over 'C' Company. It was important for a CO to establish the fact early that he personally was unafraid, for nothing can give junior officers more confidence.

My Company Commander, John Penman, had left to become Adjutant of the 8th Argylls, and now I found myself, at the age of twenty-two, commanding my own company for the first time. I felt tough, confident and professional but had lost my earlier sympathy for the Jocks and had yet to understand the effect of the constant strain of terrorist warfare on some soldiers. This was now brought home to me with a shock. My company was guarding Lydda airport and one particular section, under a corporal, was isolated at a post on the perimeter and had not been relieved for a week.

I should have understood that they had been under great strain but the first I knew of this was to be awakened in the middle of the night by one of my Jocks who said, 'Some of the boys are bugging out.' I shouted for my Sergeant Major and together we drove across the airfield in my Jeep to find only the corporal in the tent, the others having vanished, leaving behind all their weapons and equipment. We made a guess that they had boarded a goods train bound for Egypt and we rushed to the railway marshalling yard. Clambering along the tops of the open wagons in the dark I looked into each and soon, sure enough, I saw my Jocks lying at the bottom of one of them. They came quietly, were charged with desertion and received severe sentences.

It was heart-breaking and I blamed myself for driving them too hard. It was bad leadership. But I had learned my lesson and from that day I went back to my original training to take great pains to assess the morale of soldiers under my command and ensure that they were not pushed beyond the limit of their endurance. I was filled with remorse over this incident. But now came the Ides of March and, for me, near-disaster.

With the British withdrawal only a few weeks away, fighting between Jews and Arabs was now general throughout Palestine. My company was deployed in and around the police compound by Jaffa railway station, commanding the railway lines and marshalling yards which divided the Jewish city of Tel Aviv from the Arab city of Jaffa. On this night there was heavy firing between Jews and Arabs and I expected that we might find ourselves involved in serious fighting with both sides.

It may seem an odd habit but for the whole of my military service I used to crave, in action, the solitude of being on one's own. Perhaps it was a primitive urge, like that of a jungle beast, to stalk around independently of the pack. I always found it mentally refreshing to get outside the close circle of a perimeter and view the position objectively. It seemed to give one a clearer tactical picture, too, and often in the future I was to adjust positions and re-site weapons because of what used to be called my 'Moonlight Sonatas'. There is a sense of great personal freedom, heightened in danger, to be on one's own—a sense of spiritual detachment when intuition drives out logic and one decides to do things for reasons not altogether apparent at the time but which later prove to be correct decisions. I believe that this is a sort of soldier's 'sixth sense' at work and I know from experience that it is rarely proved wrong.

But the 18th March 1948 was not such a night. At about 2 a.m., when most of Company Headquarters were asleep, I found that my signaller had received no response to his check call by wireless to the platoon I had stationed in the main station building. This was commanded by Neil Buchanan, who was later to be killed with the Regiment in Korea under circumstances of great gallantry. I therefore decided to go out and see for myself that they were all right. Foolishly, I decided not to wake my sergeant-major or batman and to go by myself for they were both tired and short of sleep. There was a lull in the firing and I walked across the railway lines, in the eerie light of arc lamps, towards the station some two hundred yards away.

Suddenly firing broke out again. I stood still for a moment, then decided that I must go on to the station and walked ahead. A terrific blow struck me and sent me spinning to the ground. It was as if my legs had been hit by a sledgehammer and I lay on the railway lines in agony, fifty yards from any cover.

What had happened was that an Argyll bren gunner on the roof of the police compound had been startled by the sudden burst of firing, seen a figure moving towards him across the railway lines and opened fire. A bullet had gone straight through my left ankle and into the right ankle, cutting the Achilles tendon of my right leg.

Woken by the firing and seeing that I had left the room which we shared, my sergeant-major, Tom Collett, ran out and saw me lying wounded. I was carried to cover and evacuated to a hospital ship lying off Haifa, where my legs were operated on by a young Australian doctor. That was the end of my stay in Palestine, only a few days before the end of the mandate.

On arrival in England I was sent to Wheatley Military Hospital near Oxford where I was told that I would probably never walk properly again. I was at Wheatley for four months and

65

greatly cheered by a visit from the Australian surgeon who had operated on me in Palestine. 'You'll be as fit as you have the guts to be,' he said, 'but you'll have a lot of arthritis in those ankles and I fear you're going to be pretty short-tempered in your old age!'

My experiences in Palestine had taught me many lessons. I now understood counter-terrorism and internal security work, both the techniques involved and the exceptional strain they impose and the consequent need to maintain morale through leadership.

I had learned the importance of maintaining friendly and close relations with the Press not only because of the effect on the battalion of the CO's difficulties, but in highly-charged political situations such as existed then, one's relations with the correspondents could sway world opinion.

I had met and made a number of Arab friends whom I liked but I had lost my romantic illusions about the Arabs and found them to be emotional, volatile, usually irresponsible and therefore responsive to firmness and a show of resolution and strength. I had, too, been immensely impressed by the Jews as enemies and decided that I would much rather have them as friends. In the three subsequent wars between Jews and Arabs, the brilliant Jewish commander General Moshe Dayan was to become one of my heroes, a man of action who combined intelligence with originality and military flair.

I had learned more about the art of leadership and about the importance of leavening toughness with sympathy. In the next two decades, these lessons learnt in a hard school were to prove their worth.

CHAPTER 5

'Men age fast on the Battlefield—and that is where I come from.'

NAPOLEON BUONAPARTE

THE WAR IN Korea now seems half-forgotten. But for me—and the other young soldiers who fought there—it was an unforgettable experience. I had had my baptism of fire in Italy. I had seen what happens when politics and soldiering merge in terrorism during the Palestine crisis. I had been twice wounded, was twenty-four years old and at the best age to face the test of being a company commander in war. In Korea, soldiers were tested to the limit. If one survived the dangers and rigours of that war, then I believe that one was truly a soldier.

My military character was tempered. Having faced this, I could face anything. There, in Korea, I achieved a self-confidence that nothing could shake.

As the Duchess of Richmond's ball provided the battle of Waterloo with a gay overture of startling contrast, so I went to war from a world of parties, balls, banquets and ceremony. When I recovered from my wounds at the Wheatley Military Hospital near Oxford, I was invited to become aide-de-camp to Lieutenant-General Sir Gordon Macmillan of Macmillan, who was General Officer Commanding-in-Chief, Scottish Command, and also Colonel of the Argyll and Sutherland Highlanders. I was very much in two minds about accepting because I dreaded 'boudoir soldiering' and in any case there were grave doubts if my wounded ankles would allow me to continue as an infantry officer. But as I limped around the grounds of the Hospital, went up to Oxford to dine in College with old friends from Whitgift and worried out the problem with myself, I decided that it was an opportunity not to be missed. The deciding factor was the high personal regard in which I held General Macmillan both as a soldier and a man. He was one of the great Divisional Commanders of the Second World War and I wanted to study

67

his approach and methods. 'Take for your model the campaigns of the great captains, that is the only way to obtain the secrets of the art of war.'

I was not to be disappointed. Not only did I thoroughly enjoy my time as ADC but I sat and listened to the great men who were his peers and acquaintances, men like Wavell, Slim, O'Connor, Wimberley and Ritchie—all great soldiers of the Second World War. It was as fascinating as meeting Wellington's generals—the Craufords and the Pagets—and it was an experience for which I remain immensely grateful.

On the military side of my job, I was kept busy. There was still a large British Army, kept up to strength by National Service, and Field-Marshal Slim, as Chief of the Imperial General Staff, had reconstituted the Territorial Army with the slogan 'Twice a Citizen'. Thus I constantly accompanied General Macmillan on visits to units in all parts of Scotland. The Edinburgh Festival was also beginning and the Army joined in with military bands and Highland dancing by ATS girls. This was a humble start to what was to become the Edinburgh Military Tattoo.

My social duties also kept me busy. After the bleak war years and their dreary aftermath, social life had suddenly blossomed again. In Scotland there was a continual round of dinner parties, cocktail parties, balls and, of course, the great games and gatherings. This was a new and exciting world, for which my childhood in the London suburbs had hardly prepared me—although I was to be grateful to that officer from the London Scottish for teaching me Highland dancing in our sitting room at Norbury.

In my discovery of this world I was helped by two great friends, Hector Macneal and David Boyle. Hector and I had served together in Italy and Palestine. His family had been given their lands in the Mull of Kintyre by Robert the Bruce and were part of Scottish history. Hector, a man of great character and zest for living, introduced me with patience and good manners to his own natural heritage of Scottish field sports—shooting, fishing and stalking—which helped me both build up my physical strength again and to recover from the endless round of parties. David Boyle was just retiring from the Argylls to farm in Perthshire. He was an Olympian character whose gallantry against the Japanese and as a prisoner of war in Malaya was legendary. Both David and Hector became keen Territorial Army soldiers and eventually commanded the 7th and 8th Argylls respectively—two splendid battalions of the famed 51st Highland Division which was eventually swept away by the military bureaucrats in Whitehall. At that time any citizen in the Territorials was living proof that the British race was still capable of giving voluntary service to the state; despite the inherent apathy of Socialism to national pride and individual enterprise.

I played my part as the dashing ADC as well as could be expected by expending with calculated profligacy a small inheritance left me by a female cousin of my father. This soon turned into a roaring open Lagonda motor car, golf at Gleneagles and romantic nights in 'Auld Reekie'. Altogether it was an enjoyable twenty-one months.

Meanwhile, the 1st Battalion of the Argylls, commanded by Lieutenant-Colonel Leslie Neilson, had gone to Hong Kong, so when fighting suddenly broke out in Korea in June, 1950, I felt certain they would become involved and was desperate to join them. General Macmillan was at first reluctant to let me go because I had not completed the customary two years as ADC, but he finally agreed on condition that I find a replacement. After frantic telephoning I found him a new ADC within a matter of hours. He took it all in very good part. So I was free to set out for my third campaign.

In retrospect, the Korean war may seem to have been very straightforward, particularly when compared with the later complexities and confusion of Vietnam. The North Koreans attacked the South Koreans and drove them and their American allies down to a small beach-head around the port of Pusan in the south of the peninsula. The United Nations then came out in support of South Korea and, under General Douglas MacArthur, massive reinforcements were poured into Pusan—at first largely American, but soon to include the first British Commonwealth troops to arrive. A counter-offensive was launched, combined with an amphibious landing at Inchon on the west coast. The North Koreans were rolled back over their own frontier, their capital, Pyongyang was captured, and their northern frontier with China on the Yalu River finally reached.

69

The war seemed won by the United Nations, when China intervened with what really were hordes. The Allies themselves were now rolled back over the frontier again, their capital, Seoul, was taken, and they were forced down into the south. A second counter-offensive was launched by the United Nations armies, now including men from sixteen nations, the British Commonwealth providing a division. Again the Communists were pushed back and the fighting finally stabilized along the approximate line of the old frontier on the 38th Parallel. There an armistice was signed in July, 1953, and talks on a final peace settlement were still droning on in 1959.

Clear-cut and dramatic as this campaign may now seem, I remember it as almost total chaos. The Argylls were in urgent need of reinforcements and in the scramble to take off for the Far East inoculations were either overlooked or given all at once, and we finally bundled everyone we could muster into Hastings aircraft of the Royal Air Force at Lyneham in Wiltshire. It took us nearly ten days to reach Japan with frequent stops to refuel. From there we sailed in a United States Navy transport for Pusan.

At this time, the only Commonwealth forces in Korea were the two battalions of 27 Brigade, under Brigadier Coad. These were the Argylls and the Middlesex Regiment. They had been rushed from Hong Kong by warship at the end of August and had almost immediately been committed to battle. Two days before we arrived at Pusan, disaster had overtaken the Argylls. But it was disaster redeemed by courage of the highest order.

On 22nd September the Battalion moved across the Naktong River to support the left flank of a new American offensive against the North Koreans. On the 23rd they attacked Hill 282. Fighting took place in the thick scrub and the evacuation of the Argyll wounded was a difficult problem. Major Kenneth Muir, second-in-command of the Battalion, went onto the hill to organize their evacuation and the North Koreans counter-attacked two Argyll companies. Ammunition began to run out and unfortunately there was neither effective tank nor artillery support. An air strike was called for. The Allies had complete command of the air and in napalm, or jellied petrol fire bombs, held a deadly weapon against infantry attack. The target was clearly described and the Argylls surrounded their own positions

70

with recognition panels to avoid any mistake by the pilots. Soon after, Mustang fighter-bombers appeared, circled, then dived and dropped napalm bombs and opened cannon fire—on the Argylls!

The hill-top erupted in flame. Ammunition began exploding. All those not killed or wounded flung themselves down the steep hillside to another position some fifty feet below the crest of Hill 282.

Kenny Muir, one of the survivors, decided that the charred and smouldering hill-top must be immediately re-occupied. Collecting about thirty Argylls he led them with a cheer back up on to that terrible hill, under heavy enemy fire. Again the hill belonged to the Argylls but now there were only fourteen men left to defend it. Hugely outnumbered and under heavy fire and with their ammunition running out, the small force fought on. When his own ammunition was spent, Kenny took over a two-inch mortar and was firing it, still shouting encouragement and advice until finally he was hit by two bursts of automatic fire, mortally wounded. His last words were, 'The Gooks will never drive the Argylls off this hill.'

For this action of Hill 282—a name that will always be remembered by Argylls—Major Kenneth Muir was awarded a posthumous Victoria Cross and the American Silver Star. He was thirty-eight years of age when he fell. His father had commanded the 1st Argylls in 1923 and the London *Evening News* said:

GALLANT TRADITION

Certain considerations of great importance emerge from the story of the award of a posthumous V.C. to MAJOR KENNETH MUIR for gallantry in Korea. They are worth the close attention of those glib and plausible people who in the years since 1945 have done their best to tamper with the deep-rooted traditions and conventions of the armed forces of the Crown.

The Korean conflict seemed, at the outset, a new kind of war, a war undertaken by the United Nations in defence of a principle without which that organization is valueless. It has turned out—in its actual waging—to be a very old-fashioned sort of war indeed.

The theorists held that the day of the infantryman was over; all wars of the future were to be fought between highly skilled, remote-control technicians. The theorists, and those with a passion for

71

administrative tidiness, held that the infantry battalion, with its great simple pride of regimental tradition, was hopelessly out of date.

Officers and men must be interchangeable, like cogs. The social theorists had their finger in the pie, too; they particularly resented the idea of family connection with a regiment—it smelt of 'privilege', of 'reaction', of social snobbery.

The Korean war has called for infantry, with the dogged, disciplined, self-sacrificial qualities of their kind. That was a lesson which was being ruefully learned when the first British detachments went into the line; and our men have never ceased to practise those simple, fundamental, soldierly virtues.

It is to be noted that MAJOR MUIR rallied and inspired his men, not with the code number of an anonymous group, and not with some such cry as 'Citizen soldiers of the social democracies of the West' but with the name of their regiment, the Argylls.

It is further to be noted that MAJOR MUIR was the son of a former commanding officer of the regiment; his life and service, his courage and his honour were dedicated to the service of a regiment in whose traditions he had been bred since early boyhood. He represented and embodied the faith—so sneered at in progressive circles—that an inheritance of this kind is a great privilege, linked inescapably with an austere duty.

It was by men like MAJOR MUIR that Britain's greatness was made; in spite of all the growing, destructive efforts of the progressives for two generations it is impossible to say that that greatness is gone when there are still men like MAJOR MUIR fighting and dying for their country.

The Argylls' casualties on that day were seventeen killed or missing and seventy-six wounded, leaving the battalion with only two rifle companies.

Back in the Battalion, I was immediately among friends. John Penman was not there, he had been wounded on Hill 282 but soon rejoined. I went to 'B' Company as second-in-command to Alastair Gordon-Ingram. John Slim was now Adjutant and I saw a host of familiar faces from Palestine. As always in a good Regiment, it is on occasions like these that the togetherness of the family spirit settles reinforcements down quickly and painlessly. This vital factor, so easily overlooked by tidy-minded Whitehall planners, is one of the strengths of the Regimental system. Had we been joining a strange unit without our bonds

72

and traditions, goodness knows what our feelings would have been.

The day before, on my way up to the line, I had met a young Australian lieutenant-colonel called Green who was joining 27th British Commonwealth Brigade with his 3rd Battalion of the Royal Australian Regiment. He was to be killed a few weeks later. We had a long talk together, and he told me that his men were all volunteers and mostly veterans of the Pacific. Having volunteered for adventure they were longing to get into the fight. Those I met struck me as being superb specimens, full of fighting spirit and utterly relaxed about life. I felt that if we were going to fight beside these warriors we were fortunate indeed.

We had arrived at the right moment, for General MacArthur had seen the quality of the British troops and had decided that they were to form the spearhead of his northward thrust. We were therefore flown from Kumchon to Kumpo airfield near Seoul, a town we were to know well in the future. At last the two British battalions were formed into a real brigade with the addition of the Australians, gunners and an Indian field ambulance. John Penman rejoined us from Japan. He had just been awarded a bar to his Military Cross and was in no fit state to be back on active service but his indomitable spirit had triumphed over hospital rules and regulations and he had bluffed his way out of Japan as 'a special emissary from General MacArthur to the Scots'!

We came under command of the 1st United States Cavalry Division for the general advance northwards. They were already through Seoul and crossing the Imjin River. But their tactics were to go barrelling along the roads in the best tradition of General Patton's advance across France in 1944. To any experienced infantrymen it was obvious that no Korean army was going to be defeated on or near the roads because the hills were comparable with the North West Frontier of India. This was going to be a foot-soldier's war in the end if the enemy had any heart to resist.

It was a beautiful land of wooded mountains with deep valleys and the inevitable paddy fields. The weather was warm and the nights fresh. It was an autumn that gave little indication of the dreadful winter that lay ahead. With us were American tanks and it was an exhilarating experience to ride into action on them,

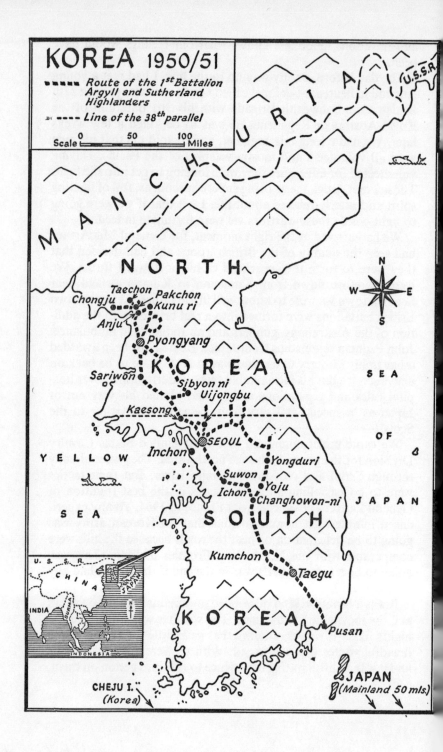

KOREA 1950/51

▰▰▰▰ *Route of the 1st Battalion Argyll and Sutherland Highlanders*

▰ ---- *Line of the 38th parallel*

Scale |‒‒‒|‒‒‒|‒‒‒| Miles
0   50   100

MANCHURIA

U.S.S.R.

NORTH

Taechon
Chongju Pakchon
         Kunu ri
Anju
Pyongyang

KOREA

Sariwon
        Sibyon ni
        Uijongbu
Kaesong

SEOUL
Inchon
              Yongduri
Suwon
Ichon   Yoju
         Changhowon-ni

SOUTH

KOREA

Kumchon
         Taegu

Pusan

YELLOW

SEA

SEA

OF

JAPAN

U.S.S.R.
CHINA
JAPAN
INDIA BURMA
INDONESIA

CHEJU I.
(Korea)

JAPAN
(Mainland 50 mls.)

blasting away at the retreating enemy with a heavy machine-gun while American aircraft swooped on to any target we gave them. This too gave a very false picture of the future.

In October, we were across the 38th Parallel and the enemy was in headlong retreat. Now came an extraordinary day; one of the most exciting in the campaign. It all happened so un-expectedly and—like our game of hide-and-seek with the German tank in the Italian farmyard—it had an air of complete unreality.

We had crossed the 38th Parallel and were still some way south of Pyongyang when on 17th October we thrust north-west to take the town of Sariwon. This was a headlong chase and none of us at my level knew what was happening on either flank. But Sariwon was an important industrial town and we expected it to be strongly defended both by the enemy retreating before us and those withdrawing from a great bulge in the coastline to the west. If they did not move quickly, these would soon be cut off by our advance.

Curiously, Sariwon appeared empty. I was moving with the leading platoon of 'B' Company and although at first there was little to be seen save bombed and burned buildings, we spotted a group of enemy, skylined on a ridge overlooking the town, and I fired at them from long range with the large 50-calibre machine-gun which the tanks carried outside their turret for anti-aircraft protection. Warily we moved in and found that it was neither held by the enemy, nor occupied by the American armour that might have got there before us from the west.

It was a strange situation and, in the centre of the town, Neil-son called an Orders Group to consider the situation and make plans for consolidating our positions in Sariwon. This confer-ence was taking place around a small group of our vehicles in the centre of the street and, as we pored over maps and wondered what was going on, we saw a large lorry driving down the street from the north towards us.

It was a large three-ton truck with high sides and it was crammed with men. We looked at them. They looked at us. Sud-denly we realized they were enemies. They were North Korean officers, who, expecting Sariwon to be occupied by their troops, had arrived to plan its defences.

For a moment we stood staring at each other. Leslie Neilson;

his second-in-command, Major John Sloane; the company commander, Alastair Gordon-Ingram; John Penman, myself and our Jocks recognized perhaps thirty enemy officers. And they, a few yards away, and with their way blocked by one of our jeeps, recognized us. Then everything happened. The CO and his party dived for shelter in a nearby grain store. Gordon-Ingram and I got behind a Jeep. Then the shooting started. There was, for five minutes, a gun battle such as the American west never knew, with Gordon-Ingram standing out in the open firing his pistol like a sheriff. Then a Jock lobbed a grenade deftly into the lorry, it exploded and there was silence. It was a brave act and showed great presence of mind but I never found out who it was. I ran from cover, jumped up on to the side of the lorry but all I could see were a dozen dead North Koreans. The rest had escaped over the back of the truck.

I looked round, the Luger pistol I had kept from the Second World War in my hand, safety-catch off. It was obvious where they had gone. A deep storm culvert ran beside the road and into a wide concrete drain. I edged towards it and saw them, huddled together in the mouth of the drain. Everyone opened fire together. It was quickly over and they all lay dead. But, in the general shooting, a small party of North Korean prisoners, standing under guard nearby, were caught in the cross fire and killed. So, after fifteen minutes, some thirty North Koreans lay dead and not one of us had been scratched.

It was an astonishing incident. Now it was to be overtaken by an even more dramatic event.

The CO decided that with the town uncleared and night falling, a firm base should be established outside the town, and he set off with a reconnaissance party to find a suitable site to the west. I took a tracked, open Bren-gun carrier and followed his Land Rover. At a small bridge, just outside the town, he stopped and waved me ahead. It was now evening and the light was failing fast. Ahead lay a narrow dusty road and down this we careered, the Colonel's Land Rover following. Then I noticed that we were, as we had half expected, about to meet other troops. Down either side of the lane files of infantry were march-towards us. They must be American. But as we drew almost level with them, I recognized them as North Koreans.

Once again recognition was mutual and instantaneous. The

officer leading one file drew his pistol and fired at me and the soldier leading the other took a shot with his rifle. Both missed. We were now bucking over the rough track between two files of armed enemy soldiers. But, after the first shots, they seemed to be too bemused to take action. I was standing up in the carrier while the driver drove on, ashen-faced, and I remember slipping the few remaining rounds of old German ammunition into the magazine of my Luger and, because there was nothing else to do and the situation suddenly seemed so ludicrous, roaring with laughter.

So my carrier, followed by the Colonel's Land Rover, followed by John Sloane and the escort, drove steadily ahead between endless files of the North Korean Army retreating from the Americans towards Sariwon. Hundreds we passed within touching distance and not another shot was fired. Some looked up at us, others just plodded on. After four miles of this inspection of the enemy we had reached their rear echelons and baggage trains and were gratified that they moved their bullock carts out of our path as we kept going.

Again I had that sense of detachment, of feeling excited, alert and amused at the same time. Doubtless, some of the enemy took us for Russians because, on that same day, another of our officers had suddenly encountered the enemy and they had asked, 'Russki?' He had replied, 'Ja, ja.' But an American officer had sworn that he was 'no goddam Russki' and been shot dead for his honesty in the first shots of a gun battle in which the Argylls were again the victors.

It was now dark and, once through the enemy column, we decided to get away from the track and take to the hills. This we did and, having concealed our vehicles under cover of a hedge, hid in a ditch until first light. We heard voices behind us but whether they belonged to enemy troops or escaping villagers I shall never know. During the night we heard the grinding of tank engines, which we assumed to be the Americans advancing behind the retreating North Koreans. Now the danger was that we could be caught by firing between the two. But, next morning, we emerged, found our vehicles intact and made contact with two surprised American soldiers drinking coffee and supposing themselves to be the most forward troops of the United Nations armies.

77

As we drove back to Sariwon I began to realise how lucky I was to have survived the last twenty-four hours. Alastair Gordon-Ingram and John Penman were waiting and were terribly relieved to see me alive. John and I had gone through a lot together. I had been best man at his wedding and we enjoyed each other's company. Little did I know that within less than three weeks both he and Alastair were to be out of the fight for good.

Sariwon was now taken and passed, but the advance continued. The 'Gooks' were on the run, the cry was 'On on'. Soon we were in Pyongyang the enemy capital, then linked up with the 187th Airborne Regiment. The next objective was Sinanju on the banks of the Chongchong river where we attacked with the Australians while later the Middlesex had a sharp battle in the hills. The resistance was stiffening as we moved on Chongju, But now the 27th British Commonwealth Brigade was halted and on 31st October 1950 we stood still within a few miles of the Yalu River and the mountains of Manchuria.

I remember feeling that this was a moment in history. But as we looked towards Communist China we were on the eve of another battle, one that was to take its place among the official Battle Honours of the Argyll and Sutherland Highlanders. Most people know the name of Balaclava and can recall the stand of the 93rd Highlanders as 'The Thin Red Line'. But few can ever have heard of Pakchon.

General Douglas MacArthur had said 'Home for Christmas' but the arrival of Chinese Communist 'volunteers' from Manchuria started what we all grimly referred to as 'a new war'. The whole strategic outlook changed. In my own mind I had never been happy that we crossed the 38th Parallel. I thought it unreasonable to expect China to tolerate us on the Manchurian border without making some effort to support the North Koreans. I was not sure what the United Nations' *aim* was to be. In addition, the Chinese electrical installations and the reservoirs in North Korea were vital to Manchuria. I also found it difficult to reconcile the astonishing effect of air power when it had been used virtually as artillery in close support of infantry. Any platoon commander could call down napalm and although it might be described as another flame-thrower I felt that the effects on the civilian population could not be overestimated. On the other hand there was ample evidence that it had helped to turn the tide and stop us being thrown out of Korea. It was one of the 'discoveries' of the Korean war but it raised grave moral problems.

We took part in several days of confused activity while the effects of Chinese intervention became apparent. We marched and counter-marched, finishing up at a place called Taechon when the battalion dug-in and I had the astonishing experience of observing massed ranks of little Chinese padding down the road towards us while the American platoon which had been deployed to stop them all lay dead in their sleeping bags—their throats cut during the night by advanced scouts who got right in amongst them. There was a popular expression 'caught in the sack' which meant exactly that.

On 5th November, Guy Fawkes night appropriately enough, we found ourselves at Pakchon. It looked much the same as when we had passed through the week before. North of the village was a bridge of sandbags on which vehicles could cross the river, but the proper crossing place, a girdered bridge, was still damaged and only passable on foot. So 'B' and 'C' Companies took up a position on the west bank of the river, the Tanyang Gang. I settled down in a small 'Gook' hut with Alastair Gordon-Ingram and John Penman and we passed a quiet night. Next morning I shaved, put my soap, towel and razor back into my small pack with my binoculars sitting on the top and walked out of the hut. Alastair, John and I had a friendly joke about the fact that 'B' Company had two seconds-in-command for the first time in history and Alastair laughingly said 'There are too many of us. It can't last.' Hardly had he spoken than firing broke out and we got orders to move back across the bridge and assemble at the little village of Kujin. Then everything happened at once. Alastair dropped with a sniper's bullet through the arm. I ran to the bridge and started hurrying the Jocks across, John Penman shouted 'Get over to the other side and sort it out'— and that was the last I saw of Alastair, my small pack, binoculars and the west side of the Tanyang Gang.

Meanwhile, back on the east bank, David Wilson, commanding 'A' Company, had been occupying a backstop position since 3rd November guarding the road to the bridge over the river Chongchong to the south. He was about five miles from Battalion Headquarters when on the morning of 5th November he

79

was ordered to clear a road-block which had sprung up between our main battalion position and himself. In fact it was an attack that had taken place against the gun lines of our supporting American Field Artillery Battalion. Fighting went on all day and 'A' Company lost five killed and six wounded until relieved by the Australians. North of this battle, we were having quite a field day ourselves, and this combined effort was to gain the Argylls 'Pakchon' as a battle honour on the Regimental Colour. As the heaviest firing was coming from behind us it was clear that we had been outflanked and that the enemy was attacking the American guns in our rear. We had therefore been ordered to re-cross the bridge, pass through our own gun lines and counter-attack.

It was clear that the nearest enemy firing was coming from a hill covered with boulders. The American tanks with us had been sprayed with machine-gun fire and their commander was refusing to join our counter-attack. I therefore took No. 5 platoon and its redoubtable sergeant, Edmundson, and advanced. We put in what was called a 'pepperpot' attack, the basis of which is fire and movement and requires much training and skill to execute properly. In essence it is that while some men dash forward, others give them covering fire before they themselves advance under cover of fire from those ahead. It appears haphazard and it is confusing to the enemy, who is offered no good targets and probably cannot even estimate the number of soldiers against him.

We reached the group of huts which were the objective and there among the rocks saw the sprawled bodies of some twenty enemy. But they were unlike any enemy I had seen before. They wore thick padded clothing, which made them look like little 'Michelin' men. I turned one body over with my foot and saw that he wore a peaked cap with a red star badge. These soldiers were Chinese. I then turned over another and, as I looked down at him, he opened one eye and looked up at me. I shot him with my Luger, shouting to the platoon, 'They're alive!' Instantly, there was a terrific gun battle among the rocks, for all these Chinese had been playing dead in the hope that we would pass through them and they could attack us from the rear. It was quickly over and all the enemy lay dead.

Reinforced with a section of the battalion machine-gun platoon and with the tanks again operational, we moved another long mile and consolidated along the line of a prominent dyke. Here we stayed while all the soft-skinned vehicles withdrew. But soon we were to see more Chinese, far more. A few hours later we were lying along the dyke, watching a scrub-covered hillside above us, when we noticed to our astonishment that the bushes, the whole hill-side in fact, seemed to be moving. It was late afternoon and obviously the Chinese were massing in their hordes for an attack once the danger of air strikes was over with the coming of darkness. I remember thinking this must surely be the end because we looked a 'thin red line' indeed. The orders came to pull back through the Australians and we went off and established ourselves in a more secure position north of Sinanju. We moved onto a hill and John Penman gave orders for the Company to form a close perimeter in readiness for an attack. By the time the platoons had moved into position it was dark and he said 'I'm just going to look around.' 'I don't believe you should, John,' I said. 'The Jocks are briefed to shoot anything that moves.' 'Not to worry,' he said and walked away from the stone sangar we were building for Company Headquarters. Almost immediately there was a shot. I ran after him shouting, 'It's the Company Commander.' He had been shot at point blank range by the nearest soldier, the bullet going into his thigh.

Weakened by previous wounds on Hill 282, this very gallant officer was to linger on for five years having been invalided out of the Army and training to be ordained as a minister of the Church of Scotland. He was a very great character and a true Argyll, who taught me a great deal about life and whose friendship I rated as a treasured possession. We carried him off the hill, a tricky game in the dark, and finally reached the Regimental Aid Post. His only concern was that I should be left in command of 'B' Company, and so it was to be for the rest of the campaign.

I became a temporary major on the 7th November, ten days before my 25th birthday. We spent the next fortnight slowly advancing again—the United Nations were having another thrust at the Yalu River. But on 25th November the Chinese launched their second offensive and this was on a much bigger scale than the previous efforts with 'volunteers'. We redeployed to Kunnu-ri and then began what the Jocks chose to call 'The Death March'—the long tramp to Sunchon, when we marched, company by company, each led by its piper, and wondered just what was happening on our flanks. I used a small map torn out of the *Daily Telegraph* to brief my soldiers on the course of the war, using scraps of information picked up regardless of authenticity. When we dug in to defend a position I would site all the major fire positions, agree the arcs for the medium machine-guns, tell the CO where we wanted mortar and artillery fire in the event of an 'SOS' and then carefully plan our escape route, mark it with stakes and take my company sergeant-major back along it. So, should we suddenly find ourselves outflanked, and this was a particular danger when South Korean troops were nearby, we could swiftly withdraw. Under these conditions, it was pointless to stand and fight for the almost inevitable result was to be over-run, as were the Gloucestershire Regiment later on in their gallant fight on the Imjin river.

Our fighting withdrawal before the Chinese advance could not always be carried out so deftly. Winter was coming on and icy winds swept down from Manchuria. We were shortly to know what it was to fight in Arctic conditions. Our uniform was all wrong—jungle green trousers, borrowed US combat jackets, bits and pieces. We marched all day to the sound of the pipes, trudging down those endless roads among the mountains and, tough as we were, the odd man began to collapse at the roadside exhausted. One of these I ordered sharply to get to his feet and march on because, 'The next man coming down the road will be Joe Stalin.' So we marched all day and at dusk prepared to face Chinese attacks. It was a grim and unforgettable experience, reminding me constantly of the fighting retreat to Corunna in the Peninsular War.

Sometimes we marched at night, too, stumbling along for perhaps twenty miles. As it grew colder it became impossible to dig in when we stopped, and we had to build low sangars of rocks. There were times when we had to leave the road and take to the mountains, carrying all our equipment, weapons, ammunition and rations on our backs. Several times were we on the point of being cut off and over-run by the Chinese corps on our heels.

Once, when the road ran through a pass in the mountains and

there was a muddled deployment at last light, I finished up with my company on the summits of some rocks, peaks 1,500 feet high. We scrambled along the razor-back ridge and the going was torture. Our wireless went off the air, it started to snow and, unknown to me, the Chinese had outflanked the whole area. I had a hurried discussion with David Wilson who was beside me with 'A' Company and the two American gunner officers who were with us. We realized from the sounds coming up from the valley and mountainside below that the Chinese had made a swift advance and were now between us and the rest of the Battalion. I believed that the only escape was a bold advance towards the Chinese, outflanking them and trying to pass through them from the rear to rejoin the other companies.

In the darkness we scrambled down the mountainside, marched round the Chinese flank, for we knew that they would be concentrated on the road, and then quietly approached the rear of the enemy's forward troops. We realized how successfully we did this when we found ourselves silently walking through their mortar lines, the crews too busy firing at the Argylls ahead to notice the Argylls passing a few yards away. Soon, by the light from a burning tank, I distinguished a familiar figure in a bonnet with an Argyll badge. It was John Slim, the adjutant, who had waited for many anxious hours to check the two missing companies through. He was his usual cheerful and encouraging self—a great son of a great soldier, Field-Marshal Lord Slim. And so we rejoined the Battalion.

Christmas 1950 was approaching and I now had a spot of luck, being detached from the battalion for a few days with my complete Company to guard the tactical HQ of the American IXth Corps at Ichon while the rest of the battalion stayed at Uijongbu. We were able to enjoy Christmas and the New Year in these plush surroundings and stocked up with food and drink. But on New Year's Day, before we had time to forget our resolutions, we were called back to Seoul to assist in the battalion task of covering the withdrawal from that city. This operation took four days and the Argylls were the last unit of the United Nations forces to withdraw across the Han River. In the eerie, deserted city, waiting for the final codeword over the wireless before withdrawing, I wondered what 1951 held in store for us.

Of the other United Nations units falling back with us at this time we had mixed opinions. The Americans could be magnificent—the American mortar company with us became almost part of the Regiment—but invariably the US Army relied too much on road transport and their minor tactics were often pathetic by British standards. The Australians we particularly liked and also the New Zealand gunners who joined the following month. We heard that the French had put up some good fights. But those for whom we had special admiration were the Turks.

I remember one day near Kunnu-ri seeing some strange-looking men marching up the road from our rear, to pass through us towards the enemy. They were tough little men wearing long greatcoats almost down to their ankles and carrying old-fashioned rifles with long bayonets fixed. This was the Turkish Brigade going into action. They fought with furious courage, and brought the enemy to a standstill with the utmost gallantry but took heavy casualties. Their medical services were meagre and so our own splendid doctor, Douglas Haldane, went across to help with the hundreds of wounded Turks who bore their suffering with extreme stoicism. There were 5,000 of them in the Turkish Brigade and they were worth three times that number.

In retrospect I suppose that the retreat was exhausting but it never struck me that morale in the Argylls should be anything other than high. Our morale held because we were the Argyll and Sutherland Highlanders and, knowing that the reputation of our Regiment depended upon our own actions as much as it had upon our predecessors', we did not crack—although occasionally people began to show signs of strain and there were moments of irritation and tension. I remember one of my Company HQ, a quiet, reliable chap, who was sitting on the ground near me, about to open a tin of baked beans, suddenly stood up and screamed. He ran wild-eyed into the middle of a paddy-field ignoring the sergeant-major's shouted orders to come back. I walked after him and he stood there waiting for me, holding the tin of beans like a grenade, shouting, 'Don't come any nearer!' I did and he flung the tin at me, but missed. When I came up to him I just told him that I knew exactly how he felt because I felt like that too, but we were all in this together and just had to stick it out or go under. At this he broke down and sobbed, walked back with me to our positions and resumed his place with the Battalion.

The greatest test was still to come. Winter had now struck Korea with all its icy horror and the danger came not so much from the Chinese, who were to suffer as much as ourselves, but from cold. For a month we held a ridge known to the Koreans as Changhowon-Ni and to the Argylls as Frostbite Ridge. With a temperature of twenty degrees *below* zero the cold was such that I could never have believed possible. If you tried to shave the shaving brush froze solid in the time it took to lift it from boiling water to your face. Anti-freeze froze. Fires could only be lit by day for fear of giving away our positions and, when the wind blew, attempts to strike matches and keep them alight with frozen fingers drove men to the depths of frustration. To add to our misery many of us developed tape worm through eating local pigs to supplement the diet of composite rations: it was animal living.

Our obsession was to try to keep warm. It was dark from about four in the afternoon, for sixteen hours, and night standing patrols had to be sent out. During daylight, the patrols would huddle round fires on the reverse slope of Frostbite Ridge keeping dry; then, when the time came for them to move forward, they would be carried on the backs of other Jocks to a point cleared of snow some two hundred yards in front of our positions so that they could start their vigil with dry feet.

Those left behind had to be constantly on the alert. We got into sleeping bags—I wore the soft, fur-lined boots I had taken from a young Chinese soldier I had killed—but I forbade anyone to use the hoods with which one's head could be covered because they impaired hearing and at night the infantryman must be like a stalking beast of the jungle with all senses alert.

We had tiny 'pup tents' which we concealed with snow behind our slit trenches and crawled into when we were not on sentry. We began to build deeper trenches to conceal small fires, carefully screened for smoke. Frostbite was a matter of good discipline and I was told later we had the lowest number of cases of any front-line unit in Korea. I only had three serious cases in 'B' Company during the whole winter. Certainly our discipline remained the 'tough but fair' traditional Argyll variety and there was no mercy shown to officer, NCO or man who failed to maintain the highest personal standards. I used to spend the best part of two hours each day on Frostbite Ridge drying out my socks and boots in preparation for the evening while Gerry Hadow, my second-in-command, kept his on to be ready if we were attacked. It was a battle of personal survival. But even without our protective hoods, we were constantly on the alert, kept awake both by the cold and by the threat of a Chinese attack.

American aircraft parachuted rations to us and, in an attempt to keep warm, I would eat two packs of concentrated rations a day. We rarely tasted alcohol except for the issue rum ration but when we did, two men would finish a bottle of whisky in ten minutes. We became like Arctic animals.

In March the weather began to ease and the counter-offensive began. Seoul was re-taken and we were moving up to the 38th Parallel through the thawing snow. One night we took up new positions along a strategically important ridge on the line of the Parallel. This had obviously been fought over before because we found dozens of slit-trenches which we would not have been able to dig in the still-hard ground. It was a dangerous place too, because that evening one of the Jocks set off a mine left from an earlier battle. But the ridge was important and having deployed the platoons and tied-in the supporting fire I settled down with Sergeant-Major Murray to keep watch beside our signaller with the wireless set who shared the slit-trench next to mine.

In the darkness I felt something move and Murray exclaimed, 'Bloody rats, sir.' I carefully switched on the shielded beam of my torch and saw that the whole trench was filled with small rats. I stood up and they were all around us. The ground was covered with thousands of the creatures, black, furry things like hamsters. They would run from the light but would swarm back over us as soon as the torch was switched off. So there was no sleep that night for anyone and we welcomed their disappearance with the dawn. But, at first light, there was another explosion— another mine had been set off, wounding an entire section under Corporal Harry Saunders who was one of my ablest and toughest NCOs.

There was something both dangerous and sinister about this ridge and as soon as it was light I went out carefully to investigate. Between two trees I saw a thin wire glistening in the dew which had formed along it. 'Trip Wires', I shouted. Then, as I looked around in the thawing snow, I saw hundreds of decaying corpses, half eaten by rats.

We had, without knowing it, deployed on an old battlefield fought over at the very beginning of the campaign. The corpses were those of both armies who had obviously been killed in and around this minefield during the initial North Korean attack. The rats had grown fat on them. I wirelessed back the state of affairs and after a couple of hours a party of American engineers came trailing up the ridge. The sergeant, a huge negro, seemed unconcerned at my warning that the place was full of anti-personnel mines and must be strewn with trip flares. His white subordinates were less confident so they left him to wander off on his own. He had not gone twenty yards when there was a dreadful bang and I turned to see the wretched man with half his head blown away. That was the end of it. The place was clearly untenable and a few hours later we were withdrawn from the position while someone in Brigade Headquarters drew a circle round it on the map and no one went near that nightmare area again.

For me personally there was then another moment of extreme danger though in somewhat bizarre circumstances. Although we always dug latrines in the company area, I have a notion for solitude in these matters and so each day would take a shovel and walk to cover some way from the company position. One morning as I was contemplating life among some rocks, I heard movement and, to my horror, saw walking warily towards me a patrol of a dozen Chinese. These were obviously extremely alert and, looking about them keenly, were coming straight at me. I managed to slip my Luger pistol slowly out of its holster and awaited the end. But although I was in no way hidden from them, they failed to see me, passing within ten yards of where I crouched. I expected them to blunder into the company positions and be shot to pieces. But they quietly veered off and slid away down the mountain. I raced back to our positions and now we saw that the patrol I had seen was being followed by a strong force of Chinese infantry, probably two hundred and fifty men. We at once called for mortar fire, but when it came the ranging was wrong and we missed what might have been our biggest kill of the war.

Later that month, holding a position on the heights in the by now marvellous spring weather, I had the task of securing a start line in the dark so that the Royal Australian Rifles could attack at first light from a secure base. As we crept quietly through the forest, with RSM Paddy Boyde bringing up porters with a load of 3-inch mortar bombs, I felt that glow of satisfaction which comes with confidence and enjoyment of the job you are doing. What better than to be a soldier on a beautiful spring night with the prospect of further adventures ahead and the companionship of the Regiment. We secured the position and as dawn came up the Australians passed through my company on the way up to their objective. Anxiously I scoured the slopes ahead with my binoculars. Benny O'Dowd, the Australian company commander and Reg Saunders, his half Aboriginal second-in-command, both old friends of mine, joked with me and passed on. But then I saw tiny figures skylined briefly against the hills and soon the attack ran into trouble. As the Australian casualties came back I went over to one badly wounded Aussie and said 'Well done Digger!' He looked up and recognized me. 'O.K. Skipper,' he said, 'I queued up for eight hours outside the Town Hall in Melbourne to get to this lot!' It was so typical of the original Commonwealth Brigade—Canadians, New Zealanders, Australians, the Middlesex and the Argylls—a marvellous fighting spirit.

In April 1951 we attacked and took yet another hill and this was to be our last. News reached us that we were to be relieved by the King's Own Scottish Borderers and return to Hong Kong. As our move began we knew that another major battle was beginning and our relief would begin their stay in Korea with a baptism of fire. But the scene was very different from what it had been in 1950.

We sailed from Inchon at the end of April, leaving behind thirty-one dead Argylls. Scores of others had been wounded and, some of these were never to recover properly. Nineteen Argylls had won decorations for gallantry—including Kenny Muir's posthumous Victoria Cross—and nine had been mentioned in despatches. But the Regiment itself had again been through the fire and survived.

As we sailed out of Inchon harbour in the US ship *Montrose*, the British cruiser *Belfast* lay nearby and, standing on a gun turret, was a lone piper, playing 'Hieland Laddie'.

It was farewell to Korea. My thoughts were very mixed. The one thing I did know was that I had come through successfully and not lost one soldier through any lack of professional competence. Perhaps we had been lucky but I knew there were times when we had created the luck for ourselves. It was all best summed up by Colonel Billy Harris of the 7th United States Cavalry, nicknamed 'The Garry Owens', who wore a yellow scarf in battle and had a regimental pride and tradition rare in the United States Army: 'Send me the Argylls! Tell the General. Tell anyone. But send me the Argylls!'

CHAPTER 6

'The strength of the British Soldier lies in his adaptability, in his obstinacy in the face of adversity, and in a sense of humour that enables him to rise above almost unbelievable hardships and difficulties. His weakness is a lack of imagination. Like the rest of mankind he has the vices of his virtues. That is why he will not take his training seriously.'

Second World War Army Training Pamphlet

WHEN I LEFT the 1st Argylls in Hong Kong at the end of 1951 I expected to be away for two years, but as it turned out I was away for six. I had reached that age when a regular officer had to gain staff experience and attend the Staff College if he was to further his career. I was sent as Adjutant to our 8th Battalion, the Argyllshire Territorial Battalion, and spent two rather lonely years living in Dunoon—a famous Clydeside holiday resort during the summer months but not much of a place for a gay young bachelor in the winter months. However, it was an ideal opportunity to study for the Staff College entrance examination and I was able to get quite a bit of shooting, though by the end of the month it was a toss-up whether I could afford to buy another 100 cartridges or take a girl friend out to dinner in Edinburgh—I usually bought 50 cartridges and took the girl to Glasgow. Fortunately I also had a very good labrador bitch called 'Texa' and got the odd shooting invitation on the strength of her nose rather than my competence with a shotgun.

At the end of 1953 I was all packed up ready to go back to the 1st Argylls when a telephone call from London told me I was going to the War Office as a General Staff Officer, 3rd Grade, in the Directorate of Military Operations. On arrival in London I found out that they dealt with the Middle East and Africa. So it turned out that I spent the next fifteen months in the War Office, passing into the Staff College in the process, learning a great deal, at 'worm's eye level' of what went on in Whitehall. When I left, the Colonel for whom I had worked said: 'You're the only sane man in Whitehall.' I took this as a twin-edged

90

compliment as he had found me sitting at my desk avidly reading Chinese philosophy when a red-faced artillery General who haunted our dreams was screaming for some futile piece of paper which I knew our clerks had failed to file.

In 1955 I went to the Staff College at Camberley. I was only twenty-nine and I think the youngest of our year. By and large it was an enjoyable experience and well worth all the sweat of passing the entrance examination. I purposely got away from the place whenever I could as I had been advised on arrival by a very sensible member of the Directing Staff that the only way to stay sane was to get out as much as possible. The end product of this excellent advice was that although I had always sworn that I would never marry until I had been to Camberley I was able to announce my engagement to Sue Phillips on our last night when we had the Staff College Ball.

My only other bit of excitement was when an Arab student, a major in the Jordanian Arab Legion, drew a pistol one night after dinner and threatened to shoot me because I was too friendly with the Israeli student who lived in the Mess. In a quick but dramatic scene I disarmed him and told him that in Britain we spoke to whom we liked. As his pistol was empty he went to bed.

From Camberley I was posted as GSO2 to the staff of the 51st Highland Division in Perth, Scotland. This was an excellent arrangement as Sue and I were able to get married in March 1956—a fact for which I am eternally grateful to the Staff College.

At the end of 1957 I returned to the 1st Battalion as a company-commander and almost immediately we were bound for active service in Cyprus. There was a ban on wives accompanying units going to Cyprus at that time but as we had no children Sue decided to fly out ahead of us and find a job as a civilian on the island. During the next eighteen months or so she was to face considerably more danger than I did but it was a wonderful experience for her and other Argyll wives followed suit.

Since 1955, the Greek Cypriots had been in revolt against British colonial rule. Although relatively small, the EOKA terrorist movement grew and came to be supported, through fear if not conviction, by the mass of the Greek Cypriots. Under its leader, Colonel Grivas, a middle-aged former officer of the Greek Army with strong right-wing views, a hard-core of terror-

91

ists was formed to carry out selective assassination and sabotage. These men, who operated in small gangs, were backed by local EOKA terrorists. They were nowhere near being as tough and efficient as their Jewish counterparts had been in Palestine, but their campaign was exceptionally ugly. It was marked particularly by the shooting in the back of unarmed men—'soft targets' as they were known—and even of women. Particularly horrible fates were reserved for Greek Cypriots suspected of helping the British. Of a total population of just under 600,000, less than a quarter were of Turkish origin and these were the only islanders upon whom the British could rely for support. As violence increasingly alienated the Greeks against the British, strong forces of the British Army and police had to be established in the island of only 3,500 square miles, until by 1958, the Governor had at his disposal more than 25,000 armed men. Even these were not enough to put down terrorism under the rules by which the game was to be played.

The campaign had reached a climax in 1956—the year in which the abortive Suez operation had been launched, partly from Cyprus—but the casualty figures were revealing. Eighty-one Britons—civilians and servicemen—had been murdered that year, but no less than one hundred and thirteen Greek-Cypriots had been killed, mostly by their fellow-countrymen on suspicion of helping the Security Forces, or of giving only grudging help to EOKA.

The following year had been relatively quiet and, although ten Britons were killed during the first three months, not one was to die from that time until the end of the year. There was optimism, even a guarded hope of a settlement, in the air when, at the end of this year, Lord Harding was replaced as Governor by Sir Hugh Foot. The Field Marshal had tackled the problem in a straightforward military fashion, basing his decisions on the edict from Whitehall that the island must be held by the British for strategic reasons and that law and order must be restored. He had had considerable success, but now, it seemed in Whitehall, a less martial figure was needed to persuade the Greek and Turkish factions to talk peace with Britain. So, in December, Sir Hugh Foot, a colonial civil servant of high repute, arrived in Nicosia.

At once Sir Hugh set about cooling the still tense atmosphere. He walked down Ledra Street—the notorious 'Murder Mile' of Nicosia—and he rode over the mountains to meet villagers with only the most discreet of escorts. A few weeks later he went to Athens with Mr. Selwyn Lloyd, the Foreign Secretary, to set in motion the intensive diplomatic bargaining between Britain, Greece and Turkey, which was to continue for more than two years.

These were the last years of trooping by sea and, in January, 1958, the 1st Argylls sailed for Cyprus in the troopship *Devonshire*. There was one stop on the way—at Algiers which was then enjoying one of the rare lulls in the war between the Arab nationalists and the French colonialists. We went ashore and one Jock who wandered into the Casbah by accident got left behind. Eventually a cruiser of the Mediterranean Fleet had to steam to Algiers to collect him—his only comment was that in French prisons they served wine with the food!

We landed in Cyprus on 1st February and were sent to the rural west side of the island. Geographically, Cyprus is divided into three main areas: a central plain stretching some sixty miles east to west; a straight, narrow chain of mountains running unbroken for about a hundred miles along the northern coast; and a *massif*—the Troodos Mountains—in the west of the island. Our area was in the north-west foothills of the Troodos where they came down to a narrow coastal plain between the little town of Polis and the twin towns of Paphos and Ktima. There had been comparatively little terrorist activity there but the rugged, sparsely-populated hills were ideal for toughening the battalion for whatever lay ahead.

At this time our main operational task was to impress ourselves on the area, 'showing the flag' in remote villages and letting the Greek Cypriots see that we were alert and ready for whatever the terrorists might attempt. It was a healthy life, living out in the hills, but there was an eeriness about Cyprus which was sometimes brought home to us by some small incident. A flashing signal light might be seen at night across a valley and we would know that there were no British troops there. And one night, my company was practising the technique of laying ambushes when into their midst walked not, as they had expected, their instructor but an unknown man who at once took to his heels.

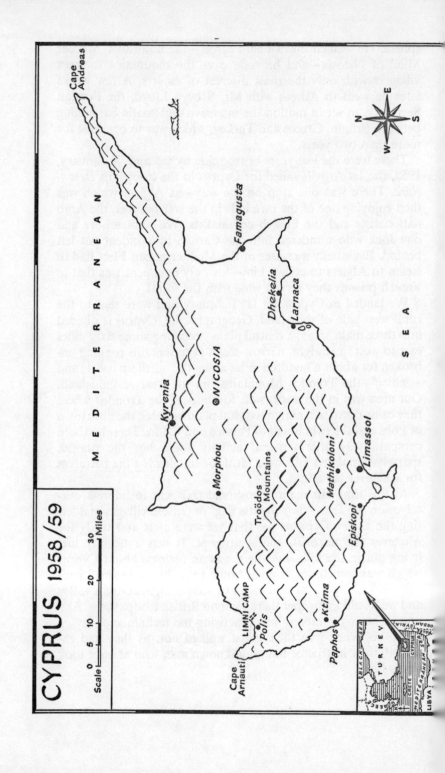

CYPRUS 1958/59

Scale

0 5 10   20   30
            Miles

So politically explosive had the Cyprus situation become that we soon learned to be extremely wary of both the civil authorities and politicians. A minor but typical example of the trouble that could be caused by soldiers obeying instructions laid down by civilians took place in a small village near our camp.

Alastair Campbell—one of my company subalterns, son of Brigadier Lorne Campbell who had won the Victoria Cross commanding the 7th Argylls in the Western Desert, took his platoon into this village to investigate reports that banned EOKA slogans had been painted on the wall of the church. He tactfully but firmly told the headman that these were illegal and must therefore be removed. He would have to order his men to paint them out, so defacing the outside of the church further. He returned later in the day to find that, far from having been removed, yet more slogans had been added. Again he warned the headman saying that this was his last chance. Early that evening, Alastair returned a second time. As his platoon entered the village, doors slammed and slops were thrown at the Jocks from upper windows. In the centre of the village, opposite the church, which was still defiantly covered with the offending slogans, the coffee shop was crowded with men waiting to see what the British would do.

They did not have long to wait. Alastair ordered his platoon to go round the shops and collect every sort of paint, ink and creosote they could find and mix it together in buckets. This done, they proceeded to cover the outside of the church with this murky mess. This was, of course, the only way this particular regulation, laid down by the Colonial Government, could be enforced.

The first I knew of the incident was when I was summoned to explain the behaviour of my men and shown a sheaf of cuttings from Greek newspapers virulently accusing British soldiers of desecrating a Greek church. The Jocks, with typical relish, had not only painted out the slogans but added a few of their own like 'Home Rule for Scotland' and 'Rangers for the Cup'.

A greater danger to us than terrorists were forest fires. As the hot summer wore on these became an increasing risk. They were particularly liable to break out in the month between the hop and grape harvests. Villagers would be paid about half-a-crown a day for fire-fighting so the starting and fighting of fires had

95

become a source of employment, on one occasion at least the fire-fighters arriving on the scene of the blaze before it had started.

We also knew the fires were started by terrorists to cover their escape from a British cordon or to trap British troops. When a stiff breeze was blowing forest fires could race up a hillside at thirty miles an hour and on one occasion an entire platoon of another unit had been caught by a fire and wiped out. One of our own platoons nearly suffered the same fate but luckily the fire was only just beginning and, with presence of mind, they ran towards and through the narrow belt of fire instead of trying to run away from it.

Sue, who had arrived in Cyprus before us, had got a job as personal assistant to Norman Beresford, British District Commissioner in Paphos, where she lived in a charming little Greek hotel. As I was often able to stay there with her, she was known to be the wife of a British officer and many of us worried about her safety. That she was unharmed was largely due to her own coolness and common sense.

In May 1958 we took part in an operation called 'Kingfisher' which had as its aim the capture of a high grade terrorist group reported to be hiding in the southern foothills of the Troodos mountains to the west of Limassol. They were thought to include Grivas himself with all his personal staff. The operation was mounted with great secrecy and initially carried out by at least three battalions. No reconnaissance was possible. The ground was hard going with precipitous ridges and valleys, strewn with scrub and boulders between carob trees and olive groves. The tactics we employed consisted of placing a wide cordon round an area of nine square miles in the initial phase, so that in theory no terrorist could slip out. Our aim was either to catch them trying to break out of the cordon at night or, by day, to go inside the cordon and look for them by searching for their 'hides'.

Grivas has written in his memoirs that he had plenty of warning of our intentions and no difficulty in escaping the slowly-closing net. I have every reason to believe him. As long columns of British Army lorries converged on the Limassol area—ourselves having to make the long journey round the west coast of the island—it was obvious to all that some major operation was afoot. Then helicopters flying observation teams to hilltops over-looking the scene of the search gave further warning and, indeed, one pilot reported seeing a body of men escaping from the area over the hills.

It took the brigade about twenty-four hours to get its cordon into position and this was too stretched to be effective even if our quarry were still within it. My own company had to cover a front of four thousand yards in hills covered with rocks, scrub and crevices. After a quiet first night, the troops descended into the bowl in the hills where Grivas was still thought to be hiding and began intensive searching, finding nothing.

But that night things did happen and we were narrowly to miss some success. It was very dark but to create a form of arti-ficial moonlight a searchlight battery was shone on to the low clouds. This would have been helpful had not the enthusiastic crews waved their searchlights about and switched them on and off so adding moving shadows to the already difficult task of spotting lurking figures. The light was, however, fairly steady when some Argylls of the Pipes and Drums who were attached to my company and formed part of the cordon, heard two men walking along a track in their direction. The strangers were throwing small stones ahead of them—a well known trick of the terrorists when they suspected that they were near a British cordon. The pipers silently slid their weapons into the firing position. Suddenly two figures came into view, one of the pipers fired but, at that instant, the searchlights went out and the tar-gets vanished in the darkness.

We knew the two supposed enemy could not be far away and, sure enough, not long afterwards two men were heard approach-ing another sector of my company's cordon. To ensure maxi-mum vigilance throughout the night the Jocks were sited in threes, so that one man remained awake and alert while the others slept. Hearing footsteps on the hillside below, one Jock silently awoke his two comrades and they got ready to fire. Just as the enemy was obviously on the point of appearing, a soldier in another battalion across the valley fired his rifle. The bullet struck a rock beside a young NCO, waking him and pre-sumably interrupting a dream of martial glory because I was told he lurched from the ground croaking, 'Stand fast the Argylls!' Again, the startled enemy made off into the darkness.

I had no idea that all this was going on as I had established my Company Headquarters further down the hill in the centre of the company area. Suddenly, in a breathless whisper the voice of the platoon commander, a boy of deep sensitivity, came over the wireless asking if he could have permission to fire. The chap in question was not at his best as a subordinate, having left Sandhurst with a field-marshal's baton firmly in his sporran. I told him that he was the man on the spot and he must do as he thought right. (Silence is the keynote of cordon operations at night until you have a definite target, in which case you require quick and very accurate shooting.) Hardly had we stopped talking when a noise like the beginning of the third world war started and my Company Sergeant-Major and I set off up the hill in the dark, never a wise move, to see what we could do. It took us about twenty minutes to get there but the noise of shooting went on unabated. I knew that all was not well as the volume of fire was so intense. Finally, arriving into the platoon position, I met the gallant Platoon Commander in the dark and gathered that the enemy had been around and even now was hiding down the side of the valley a few feet below us.

It was one of those typical occasions when you have nothing to go on but your intuition. I had a sinking feeling that the whole thing was a mistake. But it never does to pass on your worries to those you command, unless they can solve them for you, so I asked my young Napoleon for a couple of hand grenades on the Confucian theme that if rape is inevitable one might as well lie back and enjoy it. I threw a couple of 36 grenades in front of the platoon position, got on the air to Battalion Headquarters and asked for some RAF Shackleton aircraft to come and drop flares. I settled down to await the inevitable post-mortem.

Lieutenant-Colonel 'Chippy' Anderson, who commanded the Argylls, arrived the next morning soon after first light and agreed that although 'A' Company was young, keen and only recently formed, it had too unsubtle an approach to the problems of a night cordon.

My own view was that we had failed badly now on two occasions and I blamed myself entirely. I had been too long away from regimental soldiering and had forgotten how exactly and precisely one has to brief chaps who are new to operations. Having been a company commander ten years before in Palestine and after that in Korea I had been too inclined to think that everyone else had plenty of experience. It was a lesson I never forgot. The platoon commander in question served on for a few more years and then retired into business. I gather that he still regales rapt audiences with tales of his combat experiences in Cyprus ... but then for all we knew the enemy *were* there that night and he was just unlucky. That is one of the great lessons about Internal Security operations for you rarely ever do know what really happened, even after the event. I have always had a pet theory anyway that Grivas's *Memoirs* are so inaccurate that he was not even in Cyprus for the greater part of the Emergency but fills that necessary symbol of Greek folk-lore, the gallant bandit chief who always wins in the end. When you think of it, nearly all the anti-British Colonial leaders finished up alive and well; ... a spell in prison was almost a *sine qua non* for a future Prime Minister in an emergent country.

During subsequent daylight searches, however, two promising clues were found. One was a trail of blood leading up the hill towards our cordon—but this was traced to my same subaltern with a nosebleed. The other was a blood-soaked bandage. This was to lead to a curious incident which showed the paucity of our Intelligence and the extraordinary straws at which we had to clutch in the hope of picking up a lead.

This bandage was flown to Scotland where a retired general practised a form of dowsing. He stood over a large map of Cyprus, holding the bloody bandage in one hand and in the other a thread from which was suspended a light globe, like a ping-pong ball. Slowly the globe began to swing to and fro along a particular line and this was marked on the map below. The dowser then changed his position several times and the line followed by the globe was marked on the map until several intersected. The map reference of the intersection was then sent to Cyprus and a search party sent to the spot indicated. On this occasion, the dowser pointed to a remote area within our search area where we discovered a network of caves, hitherto unknown to the British, which showed signs of recent use. Sappers were called in to blow them up so that if, in fact, EOKA had been using them as a hide it would have been for the last time.

Operation 'Kingfisher' was a failure. My own 'A' Company, which was reinforced by the Battalion Pipes and Drums and a troop of Royal Artillery, employed in the infantry role, had come in for a lot of criticism because we had apparently missed two good chances of a 'kill'. The criticism was fair even for a newly-formed company almost entirely composed of National Servicemen. But I had great faith in the company and as we had always had high morale and great spirit I had no fears that we would not profit from our mistakes. We may have lacked experience and training but Operation 'Kingfisher' supplied us with both, and from then on we were to do very much better. I was made responsible for the internal security of the twin coastal towns of Paphos and Ktima while the Artillery Regiment which normally controlled it was being relieved by the Durham Light Infantry. Both these units were to be non-operational for a month. It was just the chance I had been looking for and I was able to put into practice the technique of dominating a hostile town that I was to use on a much larger scale in Aden ten years later. The essence of this was to place observation posts and machine-gun positions on the tops of the highest or most strategically suitable buildings. These would have interlocking fields of vision and arcs of fire and, under their watchful protection, foot patrols could move about the narrow streets below in comparative safety.

Under our firm control, Paphos and Ktima were relatively quiet but the background to our lives was the atmosphere of tension that keeps the senses constantly on the alert.

According to the record at the time we induced a healthy respect for the security forces, stopped all violence and boosted the morale of the police so that they became increasingly effective in what for them as for us was a very difficult situation. Oddly enough, we also became quite popular in the town itself, the inter-factional feeling was so bitter between Greek and Turk that in their heart of hearts I believe they welcomed a strong British grip on the area. Like most ordinary people their main object was to live in peace.

It was Alastair Campbell, keen as always, who achieved our biggest success in Paphos in terms of capturing enemy and material. He was carrying out some snap road checks late one Sunday morning and feeling particularly tired as there was little rest for anyone at the time. He stopped two men on a motor cycle, carrying a sack. In the sack he discovered some small bags of crystals and powder and, underneath, a number of circular metal discs. In the centre of each was drilled a hole. They could have been plumbing equipment but they could also be baseplates for grenades. He decided to take the men in for questioning. So they were sat in the back of a Land Rover, each guarded by a Jock with his bayonet just tickling the suspects' chin. Another Jock tossed the sack into the Land Rover and sat on it and the vehicle lurched off over the rough track towards the police station.

On arrival, one of the Jocks told Alastair, 'You know, I think we have something here, sir. Those men were sweating.' 'So would you,' replied Alastair, 'if you had been sitting like that with a bayonet at your throat.' But that was not why they had been sweating. At the bottom of the sack were found eighteen sticks of dynamite in a highly volatile state, liable to blow up if roughly handled. All through the rough ride in the Land Rover, the two EOKA men—as they were proved to be—expected the entire party to be blown to pieces. I planted the experience of Paphos/Ktima into my memory as an example of what you can do in an urban area if you really dominate it and drive your own men hard so that it is an unceasing and relentless vigil over the civil population and the terrorists who hope to move freely among them. It was a seed which was to grow over the next few years, to be nurtured in Zanzibar and on Internal Security duties in East Africa and finally to bear fruit in Aden ten years later.

Experience of fighting terrorists in Palestine was proving its value to the minority of the battalion who had served there ten years earlier, but, for most, Cyprus necessitated learning new techniques. When reports of an ambush came from another part of the island, the details were analysed and we discussed how it might have been avoided and how the tables might have been turned on the enemy. We practised laying our own ambushes, putting down cordons and searching villages. The Jocks were instructed in the most practical but tactful methods of searching individuals ('Check inside hat, lining and seam band. Check all pockets of overcoat, jacket and trousers. Feel lining carefully. Be suspicious of patches of recent stitching.')

We studied the enemy's chain of command down from his Command Headquarters through to his independent groups in the countryside and the static village groups and, in the towns, the selective killer groups, opportunity killer groups, bomb-throwing groups and women's groups. But chances of a direct clash with an enemy gang were, we knew, remote. More likely was the ambush or the attack on the isolated soldier or outpost.

Although tension had been steadily mounting we had not lost one man killed—until 15th September that year. A party of Jocks were returning from an 'away' football match after dark when their truck ran into an EOKA ambush on the road. In a sharp fight two Jocks were hit and one, shot in the stomach, died before he could be given medical aid.

The Argylls reacted immediately. There might be little hope of catching the ambush party, but we cordoned and searched three villages along that road. We had carefully rehearsed such an operation, using minimum force. I have often wondered whether the Cypriots ever considered what would have happened to them had the soldiers been German, Russian or Turkish rather than British.

The experience of my company was typical. It was night when I threw a cordon round the village I had to search and myself walked quietly into its centre with my company headquarters. Then I fired a flare and the Jocks converged from all sides, hammering on doors and ordering all males into the centre of the village. There, they were asked for their identity passes and screened by the police. I had impressed upon my Jocks that there must be no brutality and that they could lay hands only on those who refused the order to move. We had no difficulty.

But elsewhere there was violence. In the village of Kathikas, another Argyll company encountered opposition. Two Greek Cypriots were killed, one attacking a Jock with a knife and another escaping from a snap check.

It was a well-conducted operation and, although we failed to catch the EOKA ambush gang, our speed of reaction stopped any further trouble.

Four days after this incident, Mrs. Barbara Castle, the Labour Member of Parliament, arrived in the island and soon enough the Greek-Cypriots were regaling her with tales of our 'brutality' at Kathikas. She visited the area at the invitation of a Greek mayor and a parade of those allegedly injured by the Argylls was held for her benefit. In a nearby detention camp she was allowed to interview detainees in the presence of the Commissioner of Paphos and an Argyll NCO. Yet she did not visit our battalion headquarters, as we expected, to hear our side of the story.

As a result of her meetings with the Greek-Cypriots, Mrs. Castle spoke to journalists about the unnecessary rough behaviour of British troops. She was quoted as saying that this was 'permitted and even encouraged' but later claimed that she had withdrawn the word 'encouraged'. But the implications were clear enough and a storm broke in the British Press and in Parliament.

Although Sir Hugh Foot had yet to be fully convinced that such allegations were untrue, he defended us as 'one of the finest regiments in the British Army' and said that he was 'not surprised' when soldiers were rough with those who resisted arrest. He added, 'When their comrades are killed, troops are naturally angry and roughness can and does take place in the heat of hot pursuit. . . .'

A visiting team from the Committee of Human Rights was also in the island and visited these villages. But now the Greek Cypriots overplayed their hand. A number of supposed casualties with bandaged heads and expressions of suffering were ordered to remove their bandages—to reveal no sign of injury. On another occasion a Cypriot was caught in the act of smashing up his own home with a sledgehammer to provide 'evidence' of Argyll vandalism.

Christopher Chataway, the Conservative Member of Parliament, sprang to our defence and was unsuccessfully sued for libel by Mrs. Castle. I personally thought the whole business was rather distasteful as I would have thought that if a British MP had got comments on the behaviour of British soldiers abroad his or her first duty was to visit the unit concerned and discuss it with the fellow-countrymen involved. An odd sequel to this story is that ten years later when the Government was debating the future of the Scottish Regiments because of the storm raised by the decision to disband the Argylls, Mr. Gordon Campbell, Conservative MP for Moray and Nairn, said of Mrs. Castle: 'I hope that she had nothing to do with this decision, although she is a member of the Cabinet.'

My own view of this and other incidents is that British soldiers are the fairest in the world when engaged on Internal

Security duties. In Cyprus there was nothing approaching the tough methods most armies would have used—ranging from torture to the shooting of hostages. The only breach of 'good manners' I recall in that particular fracas was when I ordered some Jocks to move an empty civilian car out of the middle of the road. They let off the brake and all went to the back to push; it gathered speed and took off over a cliff.

But a new situation now arose. In September 1958 we were based back at Limni Camp when it was decided that a complete rifle company of the Argylls should go to Cyrenaica to join the Royal Air Force Station at Tobruk and El Adem. Although I cannot disclose the exact nature of our duties it was at the time of a turbulent Middle East situation further aggravated by events in the Lebanon, Iraq and Jordan. My Company was detached from the Battalion and sent by sea to Tobruk. We spent the last three months of 1958 there and as Sue managed to come across too it was a welcome rest from the strain of Cyprus. We returned before Christmas and were stationed in Larnaca until the end of the Cyprus Emergency early in the following year. Although we kept up our counter-terrorist vigilance to the end the steam had been taken out of the crisis. Much credit for this goes to General Ken Darling, the last Director of Operations, who introduced a note of utter professionalism into the handling of military affairs and whose organization and methods were first class.

When our families arrived out and many soldiers were living in hired accommodation in Larnaca we had to lay down strict rules for their personal security, i.e.:

1. Never go out alone unarmed.
2. If you have to shoot, stand steady, rest your arm or take up a lying position. Never shoot on the run—you will miss.
3. Don't always stick to the same route, vary your routine.
4. Keep your eyes open when you leave your quarter; if you see anyone suspicious lurking about, challenge them at once and be ready to defend yourself.
5. Be constantly on the alert when youths on bicycles are in your vinicity.
6. If you are driving a car, keep your weapon in easy reach. If you are a passenger, have your weapon in your hand.

7. Avoid routes prone to traffic jams—don't dream; look about you and watch through your rear-view mirror.
8. Don't take a chance and answer the door at night unarmed. Have your weapon ready to hand at all times.
9. If you have to wait about in the street, get your back against a wall and look about you.
10. Don't stand about and turn your back, particularly when shopping.
11. Keep well to the side of the road, pay particular attention to corners of buildings, alleyways, doorways and shop fronts.
12. Don't ever think that because you are unimportant you are safe.

Soon it was clear that a settlement between Britain, Greece, Turkey and the Greek and Turkish Cypriots was likely and that the island was to become an independent republic in 1960.

Terrorism, counter-terrorism and internal security is demanding on the soldier, both physically and mentally. It calls for diplomacy as well as fighting qualities and a special kind of wariness and cunning. It can be a great strain on the nerves.

But occasionally there are light moments to break the tension with badly-needed laughter. One was provided by a dramatic cloak-and-dagger operation carried out by the Argylls. It was believed that the monastery of Stavrovouni was deeply implicated with the terrorists and therefore must be searched. But, set high on a mountain peak, the only approach was by a winding road which was obviously under continual observation and so would give whatever guilty men were there time to hide themselves or their weapons. So it was decided that an Argyll subaltern and three Jocks should scale the sheer cliff on the far side of the monastery from its entrance and climb over the wall to take the inmates by surprise.

The four Argylls hung sheep bells round their necks, went down on all fours and, in the night, crawled up the mountainside among the flocks of similarly tinkling sheep. Finally they reached the cliff face and, after an exhausting climb, gained the monastery wall. Weary, but confident of having achieved complete surprise, they hauled themselves up the wall—to find set out in a neat row along the top four glasses and four bottles of beer!

Our own best joke on the Cypriots came at the end of the Emergency when the island was eagerly awaiting the return of Grivas, the EOKA leader. Great mystery surrounded him. Was he in the island and, if so, where had he been hiding? Did he, indeed, exist at all? Perhaps he had died and his name been assumed by another terrorist leader? Both Cypriots and British awaited his promised re-appearance with intense interest and much imaginative speculation. In Larnaca, a rumour spread that he was about to make his first public appearance there and crowds began to gather in the streets.

It so happened that one of our company commanders, a great 'Argyll character', bore a remarkable resemblance to Grivas, stocky and fiercely moustached. He therefore dressed up like Grivas in boots, breeches, battledress top and EOKA beret. Then with the Drum Major's mace, he took his place at the head of our pipes and drums and, twirling his mace, led them through Larnaca to the stirring music of Scotland.

The Cypriots had had many bewildering experiences during the past few years but none matched the sight of Grivas making his triumphal appearance at the head of the Argyll and Sutherland Highlanders.

At last the long, sordid campaign was over. By standards of conventional war, casualties had been low but there is a peculiar horror about terrorism: the betrayals and stealthy murders, the ugly threats and the indiscriminate killing by bomb and mine. In all, 508 Britons had been killed but the total of dead Greeks, most of them killed by EOKA, had reached 1,260. And the bitterness remained to ignite in fierce inter-communal fighting between Greek-Cypriots and Turkish-Cypriots in the years ahead and, twice, to bring Europe to the brink of full-scale war.

But at first there seemed reason for optimism. All parties agreed on a constitution for an independent Cyprus within the British Commonwealth. The President would be Greek-Cypriot —Archbishop Makarios the nationalist leader—and his Vice-President Turkish, who would hold powers of veto in foreign affairs, defence and some financial questions. Of the Government ministers, seven would be Greek, three Turkish, in the House of Representatives there would be thirty-five Greeks and fifteen Turks and this proportion, geared to the balance of population, would run throughout public life. Britain, for her part, would retain two Sovereign Base Areas.

At the time it seemed miraculous that agreement had been reached and that, for each of us individually, there would now be no need to stand automatically with one's back to a wall and one's weapon loaded.

CHAPTER 7

'Ex Africa semper aliquid novi.'

PLINY THE ELDER

(Meaning that there's always something new out of Africa)

EVER SINCE I was a boy I had been fascinated by stories of soldiering on the frontiers of the Empire, of the men who carved out careers with their own swords, leading Indian soldiers, Sudanese levies, Chinese mercenaries or African askari. At Foyle's in the Charing Cross Road, before the Second World War, you could buy for 1s. second-hand editions of the biographies of these Victorian soldiers—not so much the great national figures such as Gordon, Kitchener and Roberts, but lesser, equally exciting men. I was as familiar with Kabul and the Khyber Pass as I was with Clapham and Kensington.

When India became independent in 1947, it had become increasingly obvious that opportunities for such adventurous soldiering would rapidly decrease to vanishing point and that I would have to hurry if I was to find this experience. Three times after 1945 I applied to the War Office for detached service with colonial forces, asking, in turn, to go to the Aden Protectorate Levies, the King's African Rifles and the Somaliland Scouts. But each request was refused and, in 1960, I found myself at the age of thirty-four with the 1st Argylls in the British Army of the Rhine, serving as mechanized infantry in the 20th Armoured Brigade. I was completing three years as a rifle company commander and due to go back to another staff appointment. Asked if I had a choice I wrote back to the War Office and said in official language that I would relish a spot of 'bushwhacking'. Soldiering in Germany, while essential experience, was not to my taste so I was thrilled to hear that this fourth request to serve on the shrinking frontiers had been granted and that I was to be Brigade Major of the 70th Infantry Brigade of the Kings African Rifles in Kenya.

In January 1961 when I arrived in Nairobi there was still

hope that Kenya might evolve as a truly inter-racial nation. Under British rule it had been moving in this direction and was, of course, already an inter-racial society. Despite their small numbers it was the Europeans who produced the bulk of Kenya's wealth, including eighty per cent of its agricultural exports largely grown in the 7,600 square miles of the fertile White Highlands which were farmed by Europeans, mostly of British descent.

The leaders of the European Kenyans thought that unless multi-racial political fronts were formed, the country would, on independence, be dominated by feuding African tribes, the most powerful of which were the Kikuyu, forming twenty per cent of the African population. This was the tribe that had staged the Mau Mau rebellion, which had just been put down but only after 13,000 lives had been lost, most of them African. The most forceful exponent of multi-racialism was Sir Michael Blundell, who put his idea into practice by forming the New Kenya Party. His vision of a multi-racial Kenya was given official British approval by Mr. Iain Macleod, then Colonial Secretary.

But Blundell had expected ten years in which to prepare his party and policies for independence. There was to be far less time and shortly after I arrived, in March, 1961, an election was held and it was seen that the multi-racial idea would certainly be over-whelmed by African nationalism if independence were granted.

Kenya badly needed a national leader and it became increasingly apparent that despite a proliferation of energetic politicians —notably young Tom Mboya, Secretary-General of KANU and his Communist-inspired rival Oginga Odinga—there was only one Kenyan of this stature and he, Jomo Kenyatta, was still under detention on charges of having been deeply implicated in Mau Mau. To release him now, after nine years, would be particularly difficult because only a year before the Governor of Kenya, Sir Patrick Renison, had described him publicly as the 'leader to darkness and death'. Meanwhile, as the machinery of British imperial rule ran down, the civil administration, police and the army were faced with a variety of problems, each one, to me, new and fascinating.

In Kenya, the most beautiful and desirable country in Africa, strongly contrasted and often opposed cultures met and sometimes clashed. Here the Negro Bantu from the south met the

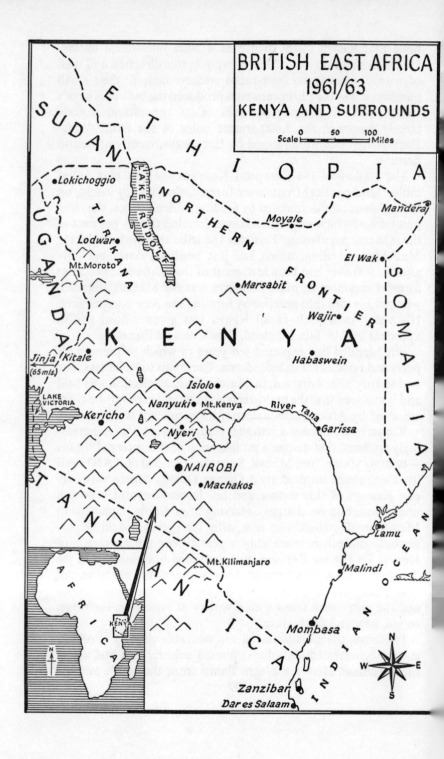

BRITISH EAST AFRICA
1961/63
KENYA AND SURROUNDS

Scale 0 50 100 Miles

SUDAN

ETHIOPIA

UGANDA

Lokichoggio

LAKE RUDOLF

NORTHERN

Moyale

Mandera

TURKANA

Lodwar

Mt.Moroto

El Wak

Marsabit

FRONTIER

SOMALIA

KENYA

Wajir

Jinja
(65 mls.)

Kitale

Habaswein

LAKE VICTORIA

Isiolo

Kericho

Nanyuki   Mt.Kenya

River Tana

Nyeri

Garissa

NAIROBI

Machakos

TANGANYICA

Lamu

Mt.Kilimanjaro

Malindi

AFRICA

KENYA

Mombasa

INDIAN OCEAN

N
W   E
S

Zanzibar

Dar es Salaam

Hamitic and Nilo-Hamitic warrior tribes. Peoples who had sprung from the Upper Nile Valley met with Muslim Somali pastoralists and pagan and Christian Ethiopians. Finally, European met African. Into the powerful tribal system of Kenya itself had come Indian immigrants while on the coast lived Arab traders, the descendants of slavers.

Brigade Headquarters of the King's African Rifles were at Nanyuki on the slopes of Mount Kenya in the lovely country which Hemingway and Ruark have written about in modern times. But it was also the land of the old white hunters in the early days of this century. Of men like Ewart Grogan who had walked from the Cape to Cairo for a romantic bet which won him his bride. Of settlers who had trekked with an ox cart and built up farms from nothing, which by hard work and enterprise were now vast areas of rich ranch land or plantation. The Kenya settlers were a race apart—tough spirited and indomitable. It was a truly wonderful country to live in and the people both black and white made it more so.

By contrast with the tough magnificence of the local settlers and the natural African scene our local pub in Nanyuki was the Mount Kenya Safari Club—a sort of Hollywood-style Gleneagles Hotel for American millionaires and film stars. But the gay social life of these last years of British rule were, to us, nothing to the excitement and beauty of the country where we lived and I was to work.

Some idea of the strangeness of it all was conveyed by the official KAR pamphlet 'Notes on Forest Lore, Operations and Training' which I browsed through on arrival. This told the novitiate what to do should he encounter lion, elephant, leopard or rhinoceros. It described in graphic military language the unpleasantness of safari ants: 'These small brown ants move in long columns on their foraging expeditions, they appear to have a well-tried system of communication which might well be studied by the majority of Regimental Signals Officers. They will crawl all over a person's body and when in position will, on a given signal, start biting together.' My imagination boggled! To an officer straight from an armoured brigade in Germany it was, to say the least, exotic.

The KAR were the descendants of a military force raised by Lord Lugard in the 70's of the last century. When I joined I

111

was met by the inevitable bunch of old hands who knew all about everything. One said 'Whatever you do, don't trust the Kikuyu, they are the Campbells of Kenya!' I pointed out that I was a Campbell myself from a Campbell Regiment and was delighted to get a Kikuyu driver and orderly from the very small percentage of that tribe who were left in the KAR since the Mau Mau emergency. They were an admirable pair in every way and, like the majority of the *askari*, loyal, industrious and excellent soldiers.

Politically it became more apparent that the European settlers were coming to the end of the road as members of the ruling power. Many of these had their capital and hopes tied up in small farms and of the European farming community of 3,500 families it was estimated some 2,000 were making about £1,500 a year or less. The future stability of the country rested on the police and the KAR and it was feared that these might break up through inter-tribal jealousies when Independence (Uhuru) eventually came.

In these final years of British rule I was extremely fortunate in being Brigade Major to Brigadier Miles Fitzalan-Howard. He was a man of astonishing vitality and enthusiasm whose leadership and drive forced through a crash programme for the commissioning of African officers which undoubtedly saved the new Kenya Army when Jomo Kenyatta took over the reins of Government. Future historians, reviewing the contribution of the British race to the advancement of mankind, may be struck by the important part which seconded British officers played in the encouragement of the emergent peoples, Miles Howard deserves a footnote to himself as he gave us all a great sense of purpose.

Soon after my arrival and before Miles arrived, the Brigade was involved in serious operations. This was on the island of Zanzibar, inhabited by Africans and Arabs in almost equal numbers, where a general election was being held. Rioting suddenly flared up and escalated into open fighting between Africans and Arabs, with appalling atrocities being committed by both sides. The stand-by company of the 5th King's African Rifles was at once ordered to the scene of action, but, before it was airborne from Nairobi, a signal reached us saying that the situation in Zanzibar was completely out of hand. So the whole of

112

the 5th Battalion was flown off, landed at Zanzibar and went straight into action. The Kenya Police General Service Units, who had preceded the soldiers, had done terrific work but the chaos was beyond them and they were exhausted when reinforcements arrived. Fighting between Africans and Arabs had been taking place, there were many dead and injured.

It took 5th KAR two days to restore order in the city and, as violence was moving out into the countryside, the 6th Battalion of the KAR was also flown in from Tanganyika. Some two thousand arrests were made and several rioters were shot. I flew to Zanzibar myself as a staff officer with Major-General Dick Goodwin, the GOC East Africa, and so had a unique overall view of the whole operation. The efficiency with which it was handled convinced me once again that firm, tough, efficient military action, if taken quickly enough, can effectively prevent a really terrible blood-letting. Methods which may appear ruthless to politicians at home are the only way of saving lives. In any case, what do politicians know of the cruel, hard facts of life when civil disorder has broken out?

Miles Howard was a great traveller, a 'swanner' as the army term puts it. We travelled the length and breadth of East Africa together from Lake Rudolf to Dar-es-Salaam and from Uganda to Lamu island—an old Arab slave port in the Indian Ocean. These safaris were quite magnificent and with our responsibility for Internal Security operations throughout the land were a combination of business with pleasure.

But our major concern was the wild, desolate Northern Province—the Northern Frontier District which began at Isiolo, on the northern side of Mount Kenya with a barrier across the track saying 'No Admittance'. This stretched for four hundred miles to the north before it met the Sudan, Ethiopia and Uganda borders and for two hundred and fifty miles to the east before it reached Somalia. This vast area was the very soul of Africa. Wild, primitive and magnificent, it was a link with the beginning of time. The people who inhabited it were even more fascinating. The Boran, the Samburu, the Turkana: magical names, primitive, pagan, wild people for whom time had stood still.

But time never stood still for the civil administrators and the Kenya Police. Peter Walters, the Provincial Commissioner and Leslie Pridgeon, the Assistant Commissioner of Police, waged a

113

constant struggle against disorder, tribal wars, cattle thieves, raiders, flood, famine, disease and drought.

The Northern Frontier District was ruled by men like these, the best sort of British colonial administrators and policemen. Like their colleagues in many parts of the former British Empire, they were men of intellect as well as of courage and extraordinary physical stamina. They lived alone in small bomas, or cantonments, hundreds of miles apart, with only a handful of tribal police to protect them. They welcomed visitors and entertained them handsomely. One of the District Commissioners in the Northern Province had a tiny swimming pool outside his house and he and any visitors would sit in it for hours in the evening, like Roman senators discussing the problems of the world. The water, in short supply until the rainy season, was purple-green in colour.

There was constant unrest along the borders of the Northern Frontier District and, from the military point of view, one of the most exciting was in the west, where Kenya marches with Uganda. Here, there had been a sort of armed 'League Football' going on across the Karamoja/Turkana border for centuries but since February 1960 and until July 1961 it had been in a state of almost continuous armed insurrection. There had been a series of raids and counter-raids between the Turkana in Kenya and the Dodoth and Karamajong in Uganda. They were fierce tribes, driven to almost constant warfare by the hopeless problem of finding sufficient grazing for their herds of cattle. In this parched, mountainous country parties of Turkana with their livestock and families would endlessly wander in search of grazing and, when times were hard, they would take to cattle-rustling. Raiding parties would try to drive away their enemies' cattle and the escaping raiders would inevitably be pursued by a war party bent on vengeance and the recapture of the stolen animals.

When such fighting was with the tribesmen's traditional weapon, the spear, casualties were few and the problem within the scope of the Kenya Police. But when I arrived in Kenya reports were coming in that the Turkana were arming with rifles. These were Austrian Steyr 8 mm. weapons with which the Italians had armed their African troops during their occupation of Ethiopia in the nineteen-thirties. They had actually been manufactured in the eighteen-nineties but they were effective weapons all the

114

same and our information was that eight hundred had reached the Turkana and were spread amongst their warriors over an area of twelve thousand square miles. To recover these rifles would be a task of immense complexity.

Operation 'Cabin Roof', as it was code-named, must have been one of the most extraordinary carried out by British commanders in the closing years of the African Empire. Seldom can a Brigade Major have had to write operation orders to include an infantry brigade, the Special Air Service and the armies and police forces of the three neighbouring countries; or in which the transport ranged from four engined RAF Beverley aircraft to thirty camels and a section of mules.

The technique used to disarm the Turkana was firm and effective but it involved no loss of life. Its success taught me, once again, that the controlled use of tough methods without brutality can prevent loss of life—the technique which I applied five years later in Aden.

When a group of Turkana were discovered we would know that their rifles were hidden and that they must be forced to hand them over. Therefore the warriors would be kept in one compound and their cattle herds in another nearby. The men were in no way molested. The headman would be asked to hand over his rifles and inevitably he would deny that he had any. During the day, the Turkana women would be allowed to bring food and water to their men but not to the cattle. Now their livestock means more to the Turkana than their wives and children and they lovingly care for these bony beasts. So when the time arrived to water the herd they would not be allowed to do so. Instead, they would be asked again to hand over their rifles and told that when that had been done they would be allowed to tend their beasts. As the hours went by, the warriors became desperate at seeing their livestock slowly wilt in the searing sun without water. Yet sometimes the Turkana would hold out for several days until, no longer able to watch their herd slowly dying before their eyes, they would tell us what we wanted to know. This technique invariably succeeded and, in this way, we collected hundreds of rifles and prevented the outbreak of serious tribal warfare.

In contrast to the type of operations carried out against the Turkana were the problems which arose on the eastern side of

115

the province because of the disputes over the ill-defined borders with Ethiopia and Somalia. In retrospect there seem to have been so many problems of the old Empire with their roots in Whitehall which the administrators on the ground, both civil and military, had to handle as well as they could. It was ever thus. The Somali Republic had been created in July 1960 when the former colonial territories of British Somaliland and Italian Somaliland were merged. The British had declared their protectorate in 1884 and in the manner of those days the territory was administered by the Government of India, then the Foreign Office, and finally, in 1905, by the Colonial Office. The population lived mainly by cattle rearing, they were pastoral nomadic tribesmen. Somalis are Sunni Muslims of Arab and not Negro stock. The nomadic migrations had traditionally ignored the artificial frontier drawn on the map. But they slowly crept south and west—the fabled 'Westward move of Islam'.

Because the border between Kenya and Somalia was a straight line, ruled north and south by some British bureaucrat earlier in the century without regard for the race and customs of the inhabitants, it both produced and perpetuated a problem. This frontier, which lay across barren scrub and desert was, of course, unmarked and so unrecognized by the Somali herdsmen who drove their herds from water-hole to water-hole in the dry season and, like their forebears, might live in either country.

So a herdsman who would be regarded as a peaceable citizen of Somalia to the east of this line ran the risk of being regarded as a bandit—what was called a Shifta—if he strayed to the west of it. Whitehall did nothing to adjust the frontier when Kenya was given independence so that all Somalis became Kenyans if they lived west of the line and, now that emergent nations jealously guard even their most barren territory in the hope of finding oil there, this particularly fatuous and dangerous British mistake has not been put to rights by the independent African rulers of Kenya.

The date for independence had been set for the end of 1963 and, as Sir Patrick Renison could hardly be expected to work easily with Kenyatta, he had resigned at the end of 1962 and was succeeded by that skilled conciliator, Mr. Malcolm MacDonald. Jomo Kenyatta became Prime Minister in May, 1963, after an election in which KANU inflicted a crushing defeat on KADU.

116

It was soon to become clear that British-style democracy would not flourish in Kenya and, in 1964, KADU was absorbed by KANU and Kenya became a one-party state.

The King's African Rifles were to become the Kenya Army, costing the new Government £2,500,000 a year. But this would be a necessity, largely because of the situation in the Northern Frontier District. Quite apart from its large Somali population the Shell Oil Company were so convinced that there was a chance of finding oil that they had begun to spend £1,000,000 a year on drilling. Somalia, which claimed much of this territory, had just concluded an arms deal worth £11,000,000 with the Russians. There was much apprehension over what might happen when the Union Jack was replaced by the Kenyan flag.

We spent much of our time in 1962 and 1963 trying to keep the peace in this troubled area because the Somalis hoped to secede on Independence and civil disobedience and rioting broke out in their main population centres of Garissa, Wajir and Isiolo. It was a fascinating time, full of unexpected and memorable experiences.

I made reconnaissance flights by helicopter over the bush country along the Tana river in search of Shifta parties. Normally we flew high so as to avoid scaring the game but on these occasions we had to fly very low and, whirring along twenty feet above the ground, I had an extraordinary view of African wild life that few can have experienced. Once, flying at this height, we suddenly came upon a large herd of elephants. A great bull, startled by the appearance of this fierce-sounding flying creature, reared up and seemed to try to catch us with his tusks. The pilot had to pull up clear of the huge beast, who stood, unafraid, while we circled and gained height, all the time his herd hurrying away to the safety of some trees. It was the most magnificent spectacle of nature I had ever watched—even better than seeing no less than three leopards resting on a rock about twenty yards from my Land Rover when out shooting guinea fowl one early morning.

Although my tour of duty in East Africa was to end about eighteen months before Kenya became independent, the 'Africanization' of the King's African Rifles was well under way. African officers were given more responsibility and encouraged to exercise the leadership which would be demanded of them

117

when we left. A number of them were most promising, but we came to realize that their methods would sometimes be very different from ours, as would be their scruples. One of our most successful officers was discovered to have achieved excellent results in a peace-keeping operation near the Uganda border by tough methods which, had we known of them, would have resulted in his immediate court-martial.

I began to realize, too, that African officers would become politicians as well as soldiers. This was dramatically illustrated when a senior African officer from Uganda visiting Nanyuki said that he planned to take over the Ugandan Army by a coup—and therefore, in effect, the country—and was I interested in becoming his Chief-of-Staff! I could, he added, name my price. I tactfully declined and have reason to be particularly glad that I did so because the coup failed and this particular officer is, so far as I know, still chained up in prison.

The three years I spent in East Africa were invaluable. The experience I gained as a staff officer was immensely varied and fascinating. I learned much about the conduct of internal security operations and of the importance of speed and firmness in such actions. In Zanzibar I got to understand more about Arabs and the Arab mind, which was to be of vital importance later in Aden. And I got to know and love one of the most beautiful countries in the world.

Relaxing one evening in the NFD after a long safari by Land Rover, I can remember sitting with an evening drink in the small KAR post at Wajir. Miles Fitzalan Howard, Pat Ross and Torquil Matheson were all laughing away about some incident during the day; the sun was sinking and with the cool evening breeze blowing across the desert, it was idyllic. I can remember thinking 'This is the best soldiering I shall ever know. Nothing could be better than this'. I think perhaps the others might have felt the same at heart because now, six years later, we have all prematurely retired—but what memories we have of those days together in the old Colonial Empire. I would not have missed it for a king's ransom.

CHAPTER 8

*'Just as tall trees are known by their shadows, so are good
men known by their enemies.'*

OLD CHINESE PROVERB

SINCE THE CAMPAIGN in Borneo ended in the summer of 1966
it has been relegated to the status of a minor, though pictures-
que, colonial war. In casualties, more than three years of sporadic fighting in the jungle seemed to have been cheap. Of one
hundred and fourteen Commonwealth dead, only sixty-four
were British. The Indonesians admitted losing five hundred and
ninety killed—many thousands short of the true figure.

The Borneo campaign was a major feat of arms by the British
Services but the full story has not yet been told. It has suited the
purpose of the British Government to play it down because with
their policy of disarmament and disengagement they could never
admit that this was a classic example of the need to have highly
trained and flexible fighting services to ensure the peace of the
world. Now that Malaysia and Indonesia, the protagonists, are
on better diplomatic terms, it is solely the demands of Whitehall
diplomacy which prevent the true lessons being drawn—that
the British have a vital part to play in the stabilization of the
free world. Therefore details of many of the operations in which
British, Gurkha, Australian and Malaysian troops, sailors and
airmen took part, together with the tactics and methods used,
remain unpublicized.

The part played by the Argyll and Sutherland Highlanders
was unique. Of all the British troops involved only they were
called upon to serve three full tours, each of six months, in the
Borneo jungles. This experience made them into a skilled and
seasoned fighting battalion in which the acceptance of responsibility became second nature. Junior NCOs did the work of
officers and platoon commanders were given wide geographical
areas which in any other campaign would have fallen to the lot
of a battalion commander. It bred self-sufficiency and confidence

119

to an astonishing degree and it soon sorted out the men from the boys.

I served only in the first of the three Argyll tours in Borneo and, as second-in-command of the battalion, had to spend most of my time on administration instead of the actual conduct of operations. But I was later to have good reason to be grateful for the Argylls' long service in the campaign and the hard work put into it by my predecessors in command, Malcolm Wallace and Glen Kelway-Bamber. It was only a few months after the end of the Borneo war that I was to lead the Battalion to Aden, knowing that nearly all of my men were already tough, battle-tested soldiers. If Waterloo was won on the playing-fields of Eton, then the battle of Crater may have been won in the jungles of Borneo.

Like so many colonial conflicts in the final years of the British Empire, that in Borneo arose out of the well-meaning efforts of British civil servants to leave their former possessions in neat and tidy order. In the early 1960s, the Federation of Malaysia had been conceived. This was to consist of Malaya, Singapore and, in Borneo, Sarawak, British North Borneo and, it was hoped, Brunei. This concept was violently opposed by Indonesia as it blocked the grandiose plans of its dictator, President Sukarno, for an Indonesian-dominated empire, intended eventually to include all these territories and the Philippines.

Both the British Government and the Malay leader, Tunku Abdul Rahman, considered the planned composition of Malaysia as essential, for it was designed to reduce the domination of the Chinese settlers in Malaya and Singapore over the native Malays. In Malaysia, there would be nearly 4,700,000 Chinese against just over 4,300,000 Malays, but this would be countered by three-quarters of a million Borneans and more than a million Indian settlers.

Now it looked as though this oddly-assorted Federation of little more than ten million citizens, so tenuously held together, would begin its life faced with the hostility of one hundred million Indonesians in the great, rich sprawl of islands which began within sight of Singapore itself.

Thus a defence treaty was signed and later confirmed between Britain and the embryonic Federation, with Australian and New Zealand support. It was hoped that with the massive British military naval and air base at Singapore President Sukarno would not dare to take direct action against Malaysia. This hope was unrealized, but the war, when it came, was to be known by the coy word 'confrontation'.

This was to be sparked off by a totally unexpected event. The little state of Brunei, with its immense oil resources, was reluctant to join Malaysia. The Sultan himself wanted to keep his state's wealth for itself and himself, but there was a strong political faction that favoured the formation of a separate new state together with the former British possessions in Borneo. Towards the end of 1962—a few months before Malaysia was to be formally established—a few well-informed British officials in Borneo warned that a rebellion might be brewing in Brunei. Their warnings were disregarded and shortly before the end of the year a coup was launched. It very nearly succeeded.

The revolt would have succeeded but for extraordinarily quick reaction by the British in Singapore. The Queen's Own Highlanders, Royal Marines and Gurkhas were rushed to Brunei, arriving only just in time to prevent the destruction of the giant oilfield at Seria and to rescue its mainly British and Dutch staff. After very little fighting, Brunei was under control and it seemed as if major trouble had been averted and that Malaysia might, after all, be established in peace.

This was not to be. Little more than three months after the failure of the Brunei revolt, the first Indonesian raiders crossed the border from Kalimantan—Indonesian Borneo—to attack elements on what was to be the Malaysian side. 'Confrontation' had begun. These attacks were stepped up. Longhouses, the immense barn-like structures in which entire villages lived in the jungle, were attacked and attempts were made to penetrate towards the towns, of which the most vulnerable was the largest, Kuching, only twenty-six miles from the Indonesian border. The earliest attacks were made by relatively small parties of Indonesian irregulars but they were known to be supported by well-armed regular Indonesian forces and it was feared that their participation would increase.

Luckily, the British had the right commander and the right troops to meet this threat. Major General Walter Walker was a Gurkha officer who had spent most of his career in the Far East, much of it on active service, and his knowledge of jungle

121

operations was unrivalled. The Brigade of Gurkhas itself had been trained for just such an operation and it was they who held the enemy while British battalions were trained and acclimatized in readiness to join them. By the summer of 1964 there were ten battalions of Gurkha and British troops in Borneo, covering a frontier running for a thousand miles along the tops of jungle-covered mountains. One of these was the 1st Argylls.

At the same time the Indonesians attempted airborne and sea-borne attacks on Malaya and Singapore itself. Again British and Gurkha troops were called for, many who were 'resting' from Borneo operations. All the raiders were killed or captured. But clearly there was worse to come—Sukarno depended on this whole concept of 'confrontation' to keep the minds of his own people off their plight at home and his crumbling regime.

During the previous year of 1963, the Argylls had been in Edinburgh, where they had been stationed on return from Germany, carrying out public duties in Scotland. These included the Queen's Guard at Balmoral and a host of activities in connection with the Edinburgh Garrison—not a particularly good training set-up for the work that lay ahead. There is no more frustrating life for a regular battalion than being at the beck and call of every staff captain who wants a fatigue squad to move a grand piano out of the NAAFI.

I cut short my tour with the KAR by six months at the request of Glen Kelway-Bamber, who was commanding the Argylls at the time. Orders had reached him to prepare for service in the Far East—the Argylls were moving again. Obviously my place was with the Regiment as 'Timber' Wood, second-in-command, was about to be promoted to command a Territorial battalion and I was needed as his relief. But with typical perplexity, the Army Department, having agreed that I could go back, immediately selected me as a student for the twenty-ninth course at the Joint Services Staff College at Latimer. Glen and I protested vigorously and said that this was a wrong decision and that I should be allowed to stay with the Argylls and go to Borneo. Reluctantly, after a lot of fuss, it was eventually agreed that my vacancy at the JSSC should be reallocated and in the words of the Military Secretary: 'A note to the effect that he has surrendered his vacancy for Regimental reasons has been put with his papers'—I was not sure whether that was to be taken as a compliment or an omen for the future. I was told that never before had an Army officer asked to be excused from pilgrimage to this Mecca of the army careerists. I think that they were rather hurt about it because the implication was that despite twenty years of 'peacetime' soldiering there were still regiments that were prepared to make rude signs to the management.

Anyway, with adventure beckoning we set out to prepare ourselves and the training period which followed must have been one of the most incongruous in the history of the British Army. We had no choice but to prepare for fighting in the dense, humid, tropical jungle of Borneo on the snow-covered hills and in the forests of Scotland in winter. So we planned Exercise 'Fiery Cross', a stimulating exercise lasting several weeks, designed not only to train the Jocks in close-quarter jungle fighting and familiarize them with the use of aircraft in a Borneo setting but also to get more recruits for the Regiment, by showing all this to the local people in our recruiting area. In both ways it was highly successful.

With assistance from Scottish Command, the Royal Air Force, the Forestry Commission and various friendly landowners, we tried to pretend that the freezing hills, lochs and forests of Scotland in November were the tropics. As our 'enemy' we were lucky to get some real live Gurkhas from the 1st Battalion 6th Queen Elizabeth's Own Gurkha Rifles who were stationed at the time in England. The Jocks and Gurkhas took to each other at once and the tough little men from Nepal and the tough little men from Scotland soon settled down together. As I drove round the training areas in mid-Argyll where my own family came from I found it utterly right and proper that a little unconventional fishing seemed to be taking place between 'battles'. Gurkhas are not the only countrymen who know how to 'guddle' trout.

The concept of the Exercise was that the rifle companies should move—usually by air—through three main training areas; on the eastern side of the Mull of Kintyre; on the western side of Loch Fyne; and, finally, in the Carron Valley west of our regimental home at Stirling Castle, where there were huge Forestry Commission plantations. It was difficult enough to imagine that these freezing woods were tropical jungle and I noted at the time, 'Undaunted, the exercise continued and frozen

123

Gurkhas flitted through the snow in brilliant moonlight, pursued by red-nosed Jocks with footprints so reminiscent of Good King Wenceslaus that the need for trackers as described in the jungle training pamphlet raised but a faint cheer of scorn.'

Yet Exercise 'Fiery Cross' was a success. It had taught the battalion to use helicopters as naturally as Land Rovers and to think in terms of close combat, patrols and ambushes. But, above all, we emerged from the wintry ordeal immensely strong and fit and eager for action.

In January 1964 I took the advance party to Singapore to take over Selarang Barracks at Changi from the Queen's Own Highlanders, whom we were to relieve. The following month the rest of the battalion arrived with their families, who were to live in Singapore for nearly three years, seeing their husbands or fathers during their spells of rest from operations in Borneo. Now we could train in real jungle across the straits in Johore and, meanwhile, I made a short reconnaissance visit to Borneo with Tom Jagger, our Quartermaster.

At the end of March, 1964, as our training programme was ending, we held a special service at the Presbyterian Church in Singapore. It was held beneath a plaque bearing the words, 'To the Everlasting Fame of the Officers and Men of the 2nd Bn. The Argyll and Sutherland Highlanders—"The Thin Red Line" —who served during the Malayan Campaign—1941-42. 244 Fell in Action. 184 Died as Prisoners of War. Their Bearing Added Lustre to the Name of Their Regiment and of Their Country'.

This was the battalion of which Field-Marshal Wavell had said: '. . . of the troops actually in Malaya I quite agree that they were insufficiently trained in jungle fighting, what that was due to I will not attempt to say. There was one battalion—a battalion of the Argyll and Sutherland Highlanders—commanded by a most remarkable Commanding Officer—which he had trained most intensively in jungle fighting. There was no doubt whatever that this battalion was as good as and better than any of the Japanese and naturally that battalion did quite magnificent work until they were practically wiped out in a battle of the Slim River on 8th January 1942 after a gallant fight.'

We were much moved by this memory of the 93rd and when,

a few days later, we set sail for Borneo we hoped that we should prove to be worthy of their great example.

The area of Borneo for which we were to be responsible was called the 4th Division of Sarawak and it consisted of over 60,000 square miles of some of the wildest country in the world. Here were primitive, headhunting tribesmen who had never before seen a white man. Here maps might misplace a mountain the size of Ben Nevis by twenty miles or fail to mark the proper course of a river as wide as the Clyde at Glasgow.

Battalion headquarters and our barracks were to be split between the small coastal towns of Miri and Lutong, where the Shell Oil Co. had sizeable establishments. Our main forward base was to be the village of Bario, 3,500 feet up in the mountains on the Indonesian border. Between the mangrove swamps along the coast and the ridge that vaguely marked the border lay dense jungle, threaded with innumerable winding rivers and creeks. The one road ran on the coast between Miri and Lutong, otherwise the only means of communication were by river, by the few jungle trails and, of course, by air.

There were two types of jungle in which we were to operate. There was primary jungle, which had never been touched by man, and here the trunks of trees would soar like the pillars of a cathedral for up to two hundred feet before opening into a canopy of leaves through which filtered dim, green daylight. Secondary jungle, as it was known, was where the great trees had been felled and had been replaced by undergrowth so dense that even a patrol of Gurkhas could only hack their way through it for half a mile in twenty-four hours.

Facing us across the border were a tough and well-trained Indonesian parachute-commando. At this stage of the campaign their attacks were on a small scale, always being made by fighting patrols of platoon strength or less. Their objectives were usually villages called longhouses, occupied by tribesmen known to be friendly to the Allies and so it was on these that our defences were based. In the Battalion's second and third tours the Indonesians would attack in company strength and more, and our positions had to be moved from the vicinity of the long-houses, which were usually in valleys by rivers or streams, into strong hilltop forts. But at this stage it was, for us, very much a platoon commander's war. A subaltern and some thirty Jocks

125

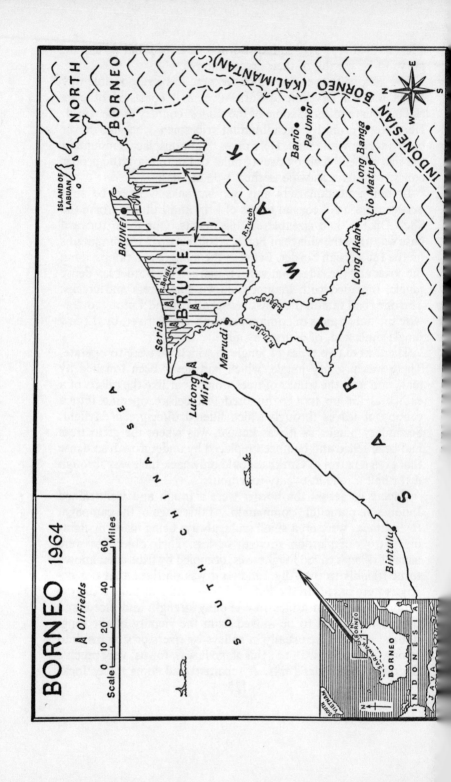

would occupy positions around a threatened longhouse—slit trenches, weapon-pits and dug-outs, surrounded by barbed wire, sharpened stakes, mines and trip-flares—and patrol the jungle from there. The nearest friendly troops would probably be twenty-odd miles away and the only communications with the outside world by wireless and helicopter. So, if there was an action, a very young officer had to accept that it was probably going to be his battle. He would have to fight it and win it himself.

This threw great responsibility on our young officers. Whereas, later, Aden was to be more dangerous to the individual, here in Borneo the subaltern risked not just taking a few casualties from sniping or grenades but, isolated in the jungle, having his command wiped out to a man. Responsibility was also thrown on the junior non-commissioned officers for, while the subaltern led a fighting patrol or was lying out in ambush one of the corporals not accompanying him might be left in charge of the defence of the position and be responsible for keeping in touch with Battalion Headquarters and seeing to re-supply by air-drop or helicopter.

This was a great and exciting challenge to the Battalion but for me as second-in-command it was personally frustrating as I was tied up in administrative duties at Miri and Lutong. Yet the problems were fascinating. The intricate system of air communication and supply was a job requiring much thought and care because this had to be run on a shoe-string. At this time, the ten thousand Allied troops in Borneo had at their disposal about the same number of helicopters as a single American company was to have in the Vietnam war.

While for those in the jungle everything was new and different, for those of us in the Battalion Headquarters there were occasional reminders that this had once been part of the British Empire. Most extraordinary of all was a Wild Man of Borneo's Henley, doubtless introduced by some British district officer who had been a rowing blue.

Each year at Marudi on the great Baram river was held the regatta which had become the biggest social event in Sarawak. While our pipes and drums played on the river bank and British and Malaysian officers and officials enjoyed cool drinks in marquees, the tribesmen raced madly in their longboats. The watching tribesmen were as enthusiastic as any crowd at Henley and

I only wished that the illusion could have been completed by the headhunters singing the Eton Boating Song. It was all very fine but I often used to sit at Lutong, near the headquarters of the civil administration and police at Miri and therefore the only place for Battalion Headquarters, and ponder on the incongruity of life there compared with that lived by the companies on the border.

In Borneo, General Walker had evolved new, indeed revolutionary, tactics for combating the invaders, tactics made possible by the helicopters and the friendly co-operation of the Dyaks, Kalabit, Iban and other friendly tribes. An operation would begin with the report of an Indonesian raiding party crossing the border, sometimes from a tribesman—perhaps one enrolled in our Border Scouts—or one of the small patrols who remained in the jungle watching the border for months at a time. The enemy would be shadowed, as far as was possible, but allowed to penetrate deeply into Malaysian territory. Then came the moment to strike. He would be ambushed and then, as he fled back along jungle trails, or occasionally by powered canoe up a river, he would again run into our 'stop' ambushes which had been placed by helicopter on all possible escape routes. Those not killed or captured might wander off into the jungle to be killed by tribesmen or die of starvation.

One of the keys to such tactics was the development of combat tracker teams, whose dog-handlers used labradors trained to follow up not only the scent of a man but of vegetation crushed under foot. In primary jungle, where the going was good, these tracker teams could move at a run after tired enemy patrols retreating laden with weapons and, perhaps, wounded.

Our first contact with the Indonesians was on 6th June 1964 but was indecisive. A platoon position at the longhouse of Pa Lungan was mortared and at once a tracker team was flown in by helicopter. But, just as the trail appeared to be getting warm, one of the Iban native trackers accidentally fired his rifle. This alerted the enemy and, as the tracker team had no support group and had lost hope of surprising the Indonesians, it retired. Later a follow-up patrol went to the area and found that the withdrawal had been prudent. A few hundred yards past the point where the team had stopped they found an enemy ambush position for upwards of sixty men. We often wondered if the

128

Iban tracker had had a sort of 'sixth' sense. Anyway, thereafter the tracker teams always went out with a support group.

This disappointment seemed, a few days later, to have been our first and last contact with the enemy. There was talk of a cease-fire and one was actually ordered. But it came to nothing and soon after fighting began again along the frontier and became fiercer in the months that followed. On the day when a cease-fire seemed a certainty, I was standing-by to go up to the frontier to negotiate with the enemy for the withdrawal of any of their troops. But then it was realized that the chance of peace had been missed and that safety-catches were to be off again. I was always sorry about this as it sounded an exciting assignment, described by our friendly and excellent Brigadier, Harry Tuzo, as 'a job for an officer who might be expendable but whom we should hate to lose!'

On 26th June, the Argylls fought their first proper action with the Indonesians. A breathless Border Scout came running into the platoon position at the Pa Umer longhouse to say that he had just escaped from a party of Indonesians who had captured his three companions and were taking them back over the border as prisoners. There followed a successful little operation directed by Lieutenant David Thomson, a most able young officer who was to be my Intelligence Officer and then my Adjutant in Aden.

In the first incident the Argylls had learned their lesson and when a tracker team was flown by helicopter to a point on the enemy's probable escape route near the frontier it was accompanied by a support group of half a platoon. The two parties crossed a major river, picked up the enemy's tracks and set off in pursuit. Torrential rain was falling and darkness was fast closing in when the tracker party climbed up from the steep bed of a narrow stream to the border. They had not known it, but in the rain and gloom, they had actually walked through the enemy's position. The support group, moving forward 400 yards behind the trackers, was ambushed and in the fire-fight that followed one Argyll was hit in the leg. In the confusion, the three captured Border Scouts escaped and joined the Argylls. One of the enemy was thought to have been wounded but, as it was nearly dark, David Thomson decided that his main objective had been achieved and he withdrew to a defensible position 300 yards back and waited until dawn. At first light, they moved forward again and found some abandoned enemy equipment but no tracks worth following. This first real contact with the enemy gave the battalion a feeling of confidence and participation in what had hitherto seemed a very quiet sector and the lessons of the action were carefully studied. For his work during the first tour in Borneo David Thomson was awarded the Military Cross.

The third and final clash with the Indonesians during our first tour in Borneo was even more dramatic. The pipes and drums were with me at Lutong and, as is the case when they feel isolated from the scene of action, I knew that they were restless and longing to be out in the jungle with the Jocks of the rifle companies. We therefore arranged for some of them to be flown to Bario and later Sandy Boswell, the company commander, sent them to a point near the Indonesian frontier to perform the strenuous but essential task of cutting a helicopter landing zone in the jungle.

So it was that at the beginning of August eight pipers under Sergeant Kenneth Robson, who three years later as Pipe Major was to play us into action in Crater, were cutting down trees in the jungle and were bivouacked there. They were interrupted by an Indonesian fighting patrol firing six 60-mm mortar bombs into Pa Lungen. As soon as the early morning mist had lifted next day two Whirlwind helicopters flew a combat tracker team, commanded by Sergeant Brian Baty, and a support group, under Sergeant Sutherland, into position and they set off in pursuit. Meanwhile a series of 'stops' were flown into ambush positions on the enemy's likely escape routes and the pipers were ordered to stop felling trees and prepare an ambush.

The tracker team caught up with the Indonesians as they were climbing the final ridge leading to the frontier. One of our Iban guides shot and killed an enemy sentry and wounded another. At once the support group put in a flank attack on the position the enemy had taken up following the first firing, drove them from it and had to kill the wounded Indonesian when he bravely continued to resist. The enemy patrol, which was now estimated to number about thirty, fled headlong towards their frontier—and ran straight into the ambush laid by the pipers. Four Indonesians were shot dead and the survivors blundered off the trail into the undergrowth, abandoning arms and equipment.

130

Apart from the six dead raiders, the Argylls recovered a rich haul of weapons—rifles, machine-pistols, mortars, bombs, grenades and ammunition—together with quantities of field equipment. What had been so remarkable about this deft little action, and what again proved the increasing self-reliance of the Argylls, was that all the sub-units involved were commanded by sergeants. Sergeant Baty was awarded the Military Medal for his part in the action and was soon after commissioned lieutenant. Three years later he was to be one of our most active officers in Aden and led a major patrol into Crater, which greatly helped me make the assessment of enemy strength and intentions leading to our successful attack.

When our six months in Borneo were up, we returned to our families in Singapore. The battalion was to serve two more such tours and, each time, they found what had been essentially a platoon commander's war had become a company commander's war. It was at this later stage that, up and down the thousand miles of frontier, were fought the fierce actions in which the Indonesians suffered so heavily, losing as many as thirty or forty killed on occasions. No wonder that when it came to the time to prepare for active service in Aden I was able to take command of the most battle-experienced battalion in the British Army. It was this, rather than any personal ability, which made the name of the Argylls a legend in Aden. The Jocks were 100 per cent professional.

CHAPTER 9

'War is a science so obscure and imperfect that custom and prejudice confirmed by ignorance are its sole foundation and support.'

MARSHAL SAXE

ONE MONTH AFTER the victory of the Labour Party in the General Election of October 1964 I found myself back in Whitehall after an absence of ten years. On promotion to brevet lieutenant-colonel in Borneo I had been 'streamed' into what the Army then called 'a small pool of highly qualified officers whose careers can be so planned that they will get experience which will best benefit the Army should they subsequently be selected for the highest ranks'. I pondered on the sort of big fish likely to be found in a small pool but was not to be disappointed, being posted as a Grade One Staff Officer to the office of the Chief of the Defence Staff. Such an appointment and the chance to see the workings of Government and the evolution of Defence policy at close quarters was a fascinating change from being second-in-command of the 1st Argylls. But it was depressing to have to leave the Battalion again because it was my only true 'home' in the Army.

Sue and I packed up our house in Singapore, paid off the two Chinese servants who brought leisure to our domestic life, and set off for Britain. Whatever else might happen we would be washing up our own dishes for the next two years. It was the tenth move we had made in nine and a half years of married life. Lorne and Angus, our two sons born in Cyprus and Kenya respectively, had now been joined in Singapore by a daughter, Colina. We were a cosmopolitan little family by any standards, more at home abroad than in London.

At this time the three fighting services numbered 423,000, the Navy (100,000), the Army (190,000), and the Air Force (133,000). To these were added civilian employees in the region of another 400,000. All were controlled by the Ministry of Defence, responsible through a Secretary of State to the Cabinet

132

and Parliament, just like any other Government department. Its size was staggering—20,000 civil servants and scientists and 3,500 uniformed staff in London alone.

I found that 'Staff of CDS' was a group of nine officers drawn from the three services and nicknamed 'The Briefers'. They fed their master, Earl Mountbatten of Burma, with succinct commentaries on the endless matters which arose in connection with his duties as Chief of Defence Staff. This seemed to require a detailed knowledge of the 'Whitehall Machine' and a host of contacts and references in the various ministries and departments with which the Defence Ministry dealt. They worked directly under Air Chief Marshal Sir Alfred Earle, Vice Chief, who had long experience of the business of Government. In addition to writing the briefs they went on tours with the CDS and prepared notes and memoranda for meetings, lectures and discussions. Like everything to do with Lord Mountbatten, they were streamlined and fast.

Mountbatten was the dominating figure in Whitehall—the last of the giants of the Second World War. His experience was vast; Chief of Combined Operations under Winston Churchill, Supreme Allied Commander in South East Asia against the Japanese, the last Viceroy of India, First Sea Lord and now in his fifth year as CDS. In March 1963 the Conservative Government had announced its plans for the reorganization of the higher direction of defence. The separate Service Ministries for the Navy, Army and Air Force were replaced by a unified Defence Ministry but it was not proposed that the three services would lose their separate identities. Each service was also to retain its professional head. These three would continue as the Chiefs of Staff Committee but under the Chairmanship of the new Chief of Defence Staff. Mountbatten had been the dynamic force behind much of this reorganization and his reputation was the subject of much jealous comment from lesser men in the 'corridors of power'. The point remains that he stood head and shoulders above them all.

I settled down to the life of the 'Briefers' which entailed a lot of hard work, usually working against the clock. Under the friendly tuition of Group Captain Denis Trotman of the RAF, who was my team leader, I confidently sent my first brief into Lord Mountbatten. By next morning it was back, covered with

133

pithy comment in his familiar green ink: 'I have wasted more time on this brief than on anything since becoming CDS!!'. I thought that this must surely be the end but was cheered up by the others. 'He's always like that with new boys,' I was told. Heartened by their encouragement I plodded on and, after a time, like the others I was writing bold and erudite comment on the endless stream of paper which came through our office.

For years I had been an ardent student of defence matters and subscribed since its inception to *Survival*, the journal of the Institute for Strategic Studies. For reasons I could never understand membership was not available to a junior officer in the Army but I noted that the political figures who had been founder members included Denis Healey, the new Labour Government's Secretary of State for Defence. When he had been the chief spokesman of the Labour Party on foreign affairs he had said, during the debate on the Conservative Government's 1961 Defence White Paper: 'The structure of our defence policy as it exists and as it is proposed in the White Paper is likely to be a major obstacle in the success of the West in disarmament negotiations with the Soviet Union.' My reaction to his speech was that if the Labour Party ever took office again we were in for disarmament and disengagement. This was the voice of 'Little England'.

From the defeat of the Spanish Armada in 1588 until the end of the nineteenth century, British strategy was based on sea power. This was dictated by geography and confirmed by commercial interest. For over 300 years it allowed a 'nation of shopkeepers' to adopt a defensive-offensive strategy entirely suited to the national goal of uninhibited self-interest. Because of it, the Royal Navy had a greater influence on the spread of western civilization than the Legions of Rome. It is astonishing to recall that Nelson's victory at Trafalgar left Britain undisputed ruler of the waves for over 100 years. It is small wonder that his dying words were 'God and my Country', for this was the embodiment of faith allied to patriotism.

This oceanic strategy, in the early nineteenth century, assured Britain's lead as a colonial power and a trading nation. The balance of power in Europe was stabilized for forty years after Waterloo and prosperity and security grew together with the

Industrial Revolution. The policy of 'splendid isolation' avoided a series of continental wars, except for the Crimean in 1854, and behind the shield of supreme naval power the high noon of Empire was reached. The British Army continued its historic role as a 'projectile fired by the Royal Navy'. Without the Navy it could never have reached its battlegrounds nor been maintained there.

The twentieth century imposed fundamental changes on British strategy beginning with the Balfour Government's Japanese Alliance of 1902 which counterbalanced any Russian expansion in Manchuria and Korea and prevented the partition of China by Russia, Germany and France at the price of holding the ring for the rise of the Japanese Empire. At the same time the decision of Germany to build a fleet in competition with the Royal Navy encouraged the growth of the *entente cordiale* between Britain and France which led to the contingency planning for British intervention with land forces if Germany attacked France. 'For the United Kingdom to attempt to play the part of a continental power is to play the fool, a thing she has been doing since 1914.' So said Major-General J. F. C. Fuller, one of the few truly great minds produced by the British Army in the twentieth century. Thus, from the beginning of this century, British strategy ceased to be oceanic.

The second fundamental change was brought about by the arrival of air power and the third by the introduction of nuclear weapons. The implications of these weapons were sufficiently misunderstood, both politically and militarily, to account for much of the artificial balance between the services. National strategy was a compromise—doomed to perpetual frustration. What added to its complexity was that new ranges of ships, aircraft and weapons became obsolescent shortly after, or even before reaching production. Costly mistakes led to loss of confidence between the public and the policy makers. Defence was the graveyard of politicians. No wonder they moved through it so quickly.

When Julius Caesar invaded Britain he was surprised to find our ancestors using chariots when the Gauls had long since discarded that weapon as being too difficult to handle in battle. The period prior to my going to the Ministry of Defence had seen the abandonment of the 'Blue Streak' and 'Blue Water'

135

missile projects; the change from 'Skybolt' to 'Polaris' and the political transformation of the projected TSR2 aeroplane from an Army support aircraft to a strategic nuclear strike aircraft. Certainly no latter-day Caesar would be able to accuse our defence planners of failing to appreciate the dangers of being caught with the wrong weapon. Nevertheless, to the Services the lack of a long-term and clearly defined strategic doctrine called for a very high degree of mental flexibility.

I believed myself that Denis Healey had the intellectual competence and the theoretical background to be a Secretary of State for Defence in the tradition of the American Robert McNamara. But I was not sure of his aim and I was deeply suspicious of any socialist defence policy because unless their own personal safety is directly threatened socialists do not like or encourage the armed services.

Nevertheless, Healey arrived in Whitehall supported by a great deal of good will from the more educated middle-piece officers in the three services, who hoped that he heralded a new era. For years past the Defence Minister had been a bird of passage. What was needed was the continuity of an expert at the top who could make the civilian and military bureaucracies work in harmony under sympathetic political direction. Defence had become an exact science. In America the writings of people like Henry Kissinger, Herman Kahn and the RAND Corporations' Research Council had opened the eyes of science and industry to the vast field of technical development which could be paid for indirectly through the national defence budget. It was 'big business'! Studies of weapons systems and military equipments based on their 'cost effectiveness' had come to dominate budget decisions so there was confidence in the Pentagon that these were the right decisions. But in the Ministry of Defence experience was to show that the feeling of confidence was not justified; human judgement would always be needed no matter how scientifically one reduced and clarified the area in which it must be exercised. The techniques of gaming, of operations research and of systems analysis, which the Americans applied to questions of military strategy, were slowly seeping into the British defence organization. Again it was largely the drive and enthusiasm of Lord Mountbatten and the Chief Scientific Adviser to the Ministry of Defence, Sir Solly Zuckerman, which encouraged these fledgelings. It looked to me as if Denis Healey had a good team to work with.

Defence planning by both major political parties was invariably better in theory than in practice. This had been one of the tragedies of our time but did not explain our national economic decline. Too often Defence was the scapegoat for other, less vulnerable fields of expenditure. The 1957 Defence White Paper had been the last attempt by Britain to maintain an independent status. It ended at the Nassau meeting in December 1962 between Harold Macmillan and President Kennedy. It was the finish of a British Defence Policy based on nuclear independence and a world-wide military posture. When the 'Blue Streak' missile was abandoned Britain could only share in the development and allotment of an American missile. After an abortive attempt to extend the useful life of the manned RAF V-bomber force into the missile age, by adapting them to launch the 'Skybolt' air-to-ground ballistic missile, this finally took the form of the 'Polaris' weapon and its nuclear submarine. These constant changes of policy had deep and disturbing effects on the three Services which was reflected in the attitude of the Chiefs of Staff, who were invariably placed in an impossible position. If they were truly objective in their decisions they risked being considered 'disloyal' to the single service to which they had devoted their lives and for whose morale and stability they were directly responsible. It was because of this that Mountbatten had always advocated more power being given to the Chief of Defence Staff, who would not shout the odds for an individual Service but give an objective opinion on the long-term national defence requirement. Human nature being what it is, this authority could never, in my view, be exercised until the Chiefs of Staff Committee was abolished. The one conclusion I came to after a further two years in Whitehall was that the overall command structure in the Ministry of Defence was wrong. It was not only wrong in its organization but was wrong in its method of selecting and appointing senior officers and in its built-in assumption that there was a continuing requirement to satisfy the needs of the existing organization rather than objectively examining ways of replacing it by a better one.

The Defence Review was only a part of the new Government's economic 'National Plan' aimed at a 25 per cent rise in the national income by 1970. In the Defence field this required stabilization of expenditure at £2,000 million by 1964 prices. One did not have to be an economist to appreciate that this particular bit of socialist theory would bring about not only a radical reduction in the forces but, thereafter, further reductions *ad infinitum*, if the only point of static expenditure in a world of rising costs and prices was to be the Defence budget. This was best expressed in a BBC satirical programme which portrayed a single, unarmed soldier in a sentry box outside Buckingham Palace bemoaning his fate as the last member of the armed forces.

The review was sold to the Chiefs of Staff and the responsible British public as a sort of confidence trick, whereby the Labour Government were going to carry out a searching analysis of British defence needs and streamline the forces to meet them in the 1970s. This particular bit of cant became all the more suspect to me when without further ado the fifth Polaris submarine was cancelled and the aircraft programme was redesigned to cancel not only the British TSR2 but also the P1154 and the HS681.

Even more sinister was that we were to buy an unproven American aircraft, the F111, to give the Royal Air Force a strike and reconnaissance capability now that there was to be no British replacement for the Canberra. Working close to some highly intelligent naval and RAF officers I could see that their concern was something more than just sour grapes to be losing what they had planned in the way of new equipments. But as a soldier I did not need any confirmation of the grave national implications of another decision, to save £20 million by reorganizing the Territorial Army, so that we were virtually to be without a home defence force or a means of reinforcing the Regular Army in a crisis.

The most charitable thing one could say about the Defence Review was that it was a classic example of muddled thinking. The proper requirement was an assessment of what strength we needed in the national interest and then a decision on how to pay for it. What we require in a democracy is a military system in which everyone believes. In Switzerland, for example, the formula is a small regular element with massive reserves. This is enthusiastically supported by the will of the nation. Sweden, a

138

socialist state more Left than anything the British Labour Party could hope for in this country, insists on developing its own aircraft and modern weapons rather than buy abroad, because this harnesses both science and industry to the national determination to be neutral but to be strong. The will of the nation stems from direct personal identification by the mass of citizens with the image of the country in the outside world. All defence efforts should therefore be based on encouraging national morale—this is the duty of political leadership.

With typical pragmatism the Government could not see beyond the fact that defence policy rests on economic circumstances and foreign policy. The former was a vote-winning factor and the latter of little interest to the majority of people in a country famed for its insularity. Nor was there any machinery like an American Congressional Committee to act as national 'watchdog', able to call for evidence from senior serving officers and civil servants if the decisions of the party in power are questionable in the long-term national interest. Parliamentary debate on defence matters in Britain is generally so pathetic and uninformed there is little wonder it becomes confined to irrelevancies.

Having got the cart well and truly before the horse the Government then began to examine the world role of the United Kingdom. If they ever had a foreign policy before the Defence Review it must have looked very different by February 1966 when Mr. Healey's Defence White Paper was presented to Parliament. Fortunately for the honour of the three Services, Admiral Sir David Luce, First Sea Lord and as such a member of the Chiefs of Staff Committee, resigned in protest. More astonishing was the resignation of Christopher Mayhew, the Navy Minister, because in his view the commitments remaining to the forces were beyond their capabilities. Healey rode through it all, though within months he was to cancel the F111 and pay compensation to the American aircraft industry for doing so and then later go back on the decision to maintain a British presence East of Suez.

My own view of all this was that a great power could still be a strong influence after it has passed the stage of Imperialism and that there was no need, even for confirmed Socialists, to act as a sort of 'moral neutralist', someone who agrees that the Western

139

Alliance must be a strong, world-wide influence but is prepared to leave it all to America and, indeed, criticize American actions in the process. I also felt that the Defence Review was trying to be too clever by half and in failing to differentiate honestly between national needs and international responsibilities it was opening us to a further decline in national character and therefore international prestige. As always, future generations would have to live with these decisions. By that I meant my own children, it was not abstract theory. It is a sad and inescapable law of life that if free men are not prepared to defend their freedom then they will lose it.

During all this time one of the most interesting aspects for me personally was the chance to watch the Civil Service at work. Whitehall was a powerful bureaucracy which ran with a motivation of its own. It was a highly disciplined organization, its frictions and conflicts rarely apparent unless highlighted by some political resignation. The civil servant of the executive class was invariably patient, intelligent and a natural fence-sitter. He did not put anything like the emphasis upon long-range planning that was displayed by his Communist counterpart, because it was his Minister who bore the responsibility for failure. These same Ministers seemed more concerned with the expediency of party political ends than with the long-term national interest. Civil servants were therefore invariably fighting a holding action, which meant in the case of military affairs that our policy was always one of containment. They were a breed of men who personified the English love of compromise. Politically, I thought from my own experience in 1954 and 1964 that they were left of centre, except when a Socialist Government was actually in power when their natural desire to 'trim' created a sort of schizophrenia and they gave astonishing advice, bearing little relation to the facts of life—for example telling the Labour Government that they could break the Rhodesian UDI in a matter of weeks by a blockade, or that South Arabia could be granted Independence under a Federal Government. Furthermore, their subservience to the Treasury and lack of business acumen made them tend to spoil every ship for a hap'orth of tar. A complete overhaul of the Civil Service was long overdue; the Fulton Report of 1968 was the first step towards it.

As the Defence Review continued I became more and more convinced that the financial ceiling imposed on Healey by the Cabinet made it impossible to arrive at a sensible military solution. All that was open to the Chiefs of Staff was a series of 'either/or' alternatives. These centred mainly around sea power and air power, but the long-term implications went further. A Navy without its carrier strike force is reduced to 'Polaris' and fishery protection; an RAF without its long-range strike and reconnaissance capability is reduced to tactical transport, shore-based maritime patrolling and firepower in support of ground forces. The next step after that could be to dispense with the RAF altogether and achieve the much discussed 'integration' of the three Services by absorbing the maritime aircraft into the Navy and the transport and ground attack aircraft into the Army. With the disappearance of the manned bomber, scheduled for the early 1970's, there would be no strategic role for the RAF and, already in late 1966, there were rumours that Mayhew's resignation pointed to the eventual abandonment of the East of Suez policy and cancellation of the F111 purchase—decisions which were announced in February 1968, two years after Healey's assurances to the contrary. This could be followed by a decision to abandon Polaris on the grounds that it was an expensive status symbol and we could finish up with a few coastal protection frigates, a continental 'cannon-fodder' type army and the final Socialist dream of no fighting services but plenty of money for domestic social services. My fears of 'Little England' were coming true.

My biggest personal worry, which was nothing to do with my job on CDS's staff, was naturally enough the future of the Army. I had come away from the Staff College eleven years earlier with a feeling of disappointment that our studies had been lacking in depth. For example the approach towards nuclear weapons was that they were just another more powerful form of artillery controlled by some dapper Gunner General. The whole question of nuclear judgement and control was ignored, except in the tactical sphere. I found this unrealistic, so indeed did some of the Directing Staff and other students. All intelligent reasoning pointed to the fact that if nuclears were introduced on the battlefield we should be engaged in all-out general war when tactical objectives were of no consequence, since nuclear battlefield influence was

141

secondary to strategic considerations and the effects on the civil population at home. Therefore the reasoning that limited nuclear war was possible must be wrong as the *consequences* could only be strategic.

The bulk of the Army was destined for employment with NATO in Western Europe if the East of Suez commitments were abandoned. Henry Kissinger had said in 1962, 'Throughout its history NATO has suffered from the difficulty that its strategic doctrine has followed, rather than guided, the creation of its forces.' It looked to me as if here again was the writing on the wall. As an infantryman I decided to 'fly a kite' to see just what the Army had in mind for its reorganization. There was still time to base this against the requirements of a world-wide military presence. Accordingly, I wrote an article 'A future for the Infantry Arm'—which was to have far-reaching effects.

The Royal United Service Institution in Whitehall was founded under Royal Charter to advance the science and literature of the three services. As a small boy I had paid my 3*d.* to visit the Museum, housed in the old Banqueting Hall on the balcony of which King Charles I had been executed. At that time I did not know that the remainder of the building housed the Lecture Room and Library of the Institution. I joined it in 1948 when I was invalided home from Palestine and for twenty years I used it to the full. Library books were despatched all around the world, and passing through London or working in Whitehall I had always found it a sanctuary where one could browse through books, attend lectures or just sit and write with every conceivable military reference to hand.

The Institution published a Journal which tended to be 'stuffy' until about 1962 when Brigadier John Stephenson, the Director, gave it a face lift. As part of the new look I sent him from Kenya an article entitled 'Life Begins at 40', which was a critical examination of the present officer career structure. For two months he argued unsuccessfully with the War Office that he should be allowed to publish it in the Journal. But they were not prepared to let what they called 'controversial' material be published by a serving officer although that very month a serving General wrote a long letter to *The Times* on the highly controversial subject of the education of the officer. I gave up in disgust, though a brother officer to whom I sent a copy had the acumen

142

to publish most of it some years later under his own name when he was at the Joint Services Staff College. Three paragraphs from my unpublished article are worth quoting still:

The serving officer fills a unique place in society. He follows a career which, except in its more technical spheres, demands the vigour and fire of youth. Unfortunately, in an effort to give prospects of employment to the age of 55, the Government sees fit to deny him any real responsibility until he is in his middle forties. This must surely be one of the longest periods of recorded apprenticeship since Moses took the Children of Israel into the Sinai Desert for 40 years.

The truth of the matter is that for the Services to guarantee employment for 35 years they must support the fiction that age and experience are, of themselves, military virtues. Historically this is nonsense when one recalls that ALEXANDER had conquered most of the known world before he died at the age of 33; NAPOLEON was 27 when he assumed command of the Army of Italy and, as recently as 1956, the brilliant Israeli attack on EGYPT was planned and led by a Commander-in-Chief, who was younger at the time than any battalion commander serving in the British Army.

The real tragedy of the present system is that although not deliberately designed to do so its effect is to hold back the younger officer until his spark has become a dull flame. The process is sometimes called 'maturing' but, unlike good wine, the aged product is all too often corked! Most officers, by the time they are 40, are contentedly running in the stereotyped grooves of a conventional career. Their family and domestic responsibilities are heavy and they are too sensibly inclined to 'play it safe' to produce the type of leadership which scaled the Heights of Abraham, won the Battle of the Nile or conceived the Chindit operations'.

Since then I made another unsuccessful attempt to publish an article, this time on African officers in the King's African Rifles, so I became used to rejection. But when in June 1966 I submitted 'A future for the Infantry Arm' I took the precaution of walking it around the 'censors' in the MOD myself and was able to get it agreed for publication in the August copy of the Journal. It was still being discussed in the Correspondence columns of the Journal two years later, so obviously it sparked off a train of thought—which was partly its intention. But a curious and unexpected by-product was that it received a degree

143

of comment in the Press and Clare Hollingworth of the *Guardian* and Ivan Rowan of the *Sunday Telegraph* wrote articles about it—while it got coverage in the Australian and Hong Kong papers. In *Punch* my suggestion that we should form a guerrilla force for the British Army, 'a hard mercenary corps, untrammelled by domestic ties and baggage, on the lines of the old French Foreign Legion' inspired Alan Coren to an amusing piece called 'The Legion of the Lost'. It also brought me a number of invitations to lecture, mostly from people interested in a permanent United Nations force for world stability. I became deeply interested in the implications of an international peace-keeping force kept on a full-time basis as an 'Army in being'.

The irony of it all was that I 'flew the kite' to try to identify the arguments against retention of the infantry regimental system (meaning, as far as I was concerned, the Argyll and Sutherland Highlanders), and at the same time argued that if any changes were going to be made towards 'rationalized, functional infantry' they must embrace the Brigade of Guards (a small force of time-expired ex-NCOs of the combatant host to play their part in the national pageantry); the Parachute Regiment (once considered the spiritual home of the progressives because of its up-to-date outlook which attracted so many forceful refugees from the antique line infantry); the Gurkhas (an un-rivalled mercenary force recruited in Nepal); and the Royal Marines (part of the Royal Navy but basically infantry). I examined them all with objective gravity and tongue well in cheek but put in my two favourite maxims 'in the final analysis recruiting is the measure of success' and 'for the infantryman his regiment, large or small, is the only basis for an *esprit de corps* which can withstand both the shock of war and the dull routine of peace. Surrounded by its traditions and nurtured on its past greatness, he achieves a sense of belonging which fills a deep psychological need and promotes the security of his personal and family welfare.'

To a number of distinguished and retired officers who during 1966 wrote to me privately saying how much they had enjoyed it I said this as part of a stock reply:

> I am a great believer in shock tactics! If the Royal Navy had woken up two years ago to the fact that they were going to have to justify their roles they would have approached the Defence Review

144

fully alive to the issues at stake. As it was they were minced up by the professional technocrats and defeated in detail.

There are few serving Infantry officers who appreciate the scrutiny to which our organizations are subjected by the new political and scientific brethren, preaching the McNamara doctrine of cost-effectiveness. I believe that if the Infantry fails to show awareness of the danger, we too shall find ourselves defeated in detail.

I very purposely took all the arguments for rationalization to the extreme of their mad, modern logic. I hope that it has produced the shock in many quarters which was my aim. The last thing I want to do is to help in the destruction of the Regimental system and I made this point several times in my article, by introducing the balance whenever I could. Nevertheless, I am convinced that the Scottish Infantry is unaware of the danger.

The complacent believe that provided we can keep up our numbers we shall survive. This is a fallacy, because it can only be true if the Infantry has a guaranteed manpower cover of something like 45,000. But what happens if it is reduced by the stroke of a pen to a mere 30,000? We shall then be faced with a panic re-organization.

I wrote these words in September 1966. By July 1968 the warning in the final paragraph had proved correct and a Supplementary Statement on Defence Policy announced that in addition to the seventeen major units being disbanded by April 1970 a further nine would be axed by September 1972. These included 1st Battalion The Argyll and Sutherland Highlanders.

CHAPTER 10

'Time spent in reconnaissance is seldom wasted'.

Military Maxims

IN THE FINAL years of the Empire successive British Governments had made great efforts to arrange their former colonies and protectorates so that these could face the future with some hopes of peace and prosperity. One formula for this was a federation of small states, geographically close but often bound together only by the chance that they had been annexed and ruled by the British since the eighteenth or nineteenth century. Such a case was the ill-fated Federation of South Arabia.

Aden itself had been an eastern Gibraltar at the mouth of the Red Sea, useful as a strategic naval base and as a fuelling station. Otherwise its uses had been so limited that it was governed from its annexation in 1839 to 1937 by officials in New Delhi on behalf of the Government of India, while the mountainous hinterland was left to its own devices unless trouble along the caravan routes called for a small military punitive expedition.

It became a Crown Colony with a Governor and Executive Council. Then, at the beginning of the 1960s, the task of welding together Aden and its hinterland into a viable federation was begun under the guidance of the then Governor, Sir Kennedy Trevaskis. In 1962, eleven states in the hinterland—feudal sultanates and sheikhdoms—were formed together into the Federation of South Arabia, and a year later they were joined by Aden itself.

It was a marriage of convenience but it was doomed from the start, unless Britain was prepared to provide stability and confidence. The hope was that 112,000 square miles of the hinterland, mostly unmapped, roadless mountains and wadis and inhabited by about one million primitive tribesmen, constantly warring amongst themselves, would settle down beside the relatively sophisticated state of Aden, of only seventy-five square miles and with nearly a quarter of a million inhabitants. The

146

two partners in the federation were quite incompatible: up country they were still living in what Europeans would regard as the Middle Ages; in Aden, Egyptian-sponsored Arab nationalist and militant left-wing trade union politics were dominant. And the Adenis felt that the British were not only intending to continue their policy of 'divide and rule' even after their promised departure but, in order to do this, were giving the up-country princelings too big a share in the government of the Federation as a whole. It was a difficult situation but no more insoluble than others, like East Africa for example, where we left reasonable stability for emergent peoples of diverse views and outlook.

At the end of 1963, as British officials and Arab rulers and politicians gathered at Khormaksar airfield to fly to London for talks on the future constitution of a fully independent Federation, a grenade was thrown at Sir Kennedy Trevaskis, fatally wounding one of his advisers, George Henderson, who gallantly threw himself between the grenade and the High Commissioner. Terrorism had started and was to grow until it reached its climax of 20th June 1967, three-and-a-half years later.

During those years of mounting terrorism I made two visits to Aden and studied the problem intently. The first was as an accompanying staff officer with Lord Mountbatten in early 1965; the second was with Air Chief Marshal Sir Alfred Earle, accompanying the Labour Government's Secretary of State for Defence, Mr. Denis Healey. These were short visits but I formed several firm opinions.

Mountbatten's visit was typically refreshing and stimulating. He had been the founding father of the tri-service joint HQ in the Middle East and, as with all his visits, it was a dynamic pageant. The unified command system in the Middle East had proved so popular with the Chiefs of Staff that it had also been extended to the Far East and Mountbatten took justifiable pride in their achievements. But I felt that perhaps there was a good deal less than met the eye in all this ballyhoo and the growth of yet another HQ superimposed on the existing single service staffs was only another manifestation of post-war Parkinson's Law. Indeed, the whole concept of having a base in Aden for intervention East of Suez carried the seed of its own destruction. South Arabia's future depended more on the outcome of the

147

Government's Review of Defence expenditure than on any sense of colonial mission. Twelve months after Lord Mountbatten's visit in 1965 the Labour Government announced that we would scuttle from Aden in early 1968.

Although since 1945 successive British Governments had conducted a policy of decolonization, at the same time they retained strategic interests in the Near, Middle and Far East which were underwritten by the presence of military bases. These bases were not necessarily at the point of decision, but being somewhere along the lines of communication they provided a military force, aircraft, stockpiles of equipment and ammunition, readily available to allow a wider strategy to be pursued.

The bases themselves depended on local goodwill if they were to survive. This point seems to have been lost on the Chiefs of Staff Committee or else they were too weak or indecisive to stand up to their political masters and point out the folly of having a base in Aden unless we were prepared to take a strong line in holding it. Military sense is entitled to prevail over political policy if the latter is obviously unrealistic and bound to lead to disaster.

Having taken the Middle East base from the Canal Zone in Egypt to Cyprus, a move involving millions of pounds, and having wasted further millions building a supplementary base for the strategic reserve in Kenya, the duty of the Chiefs of Staff lay in resisting a big build-up in Aden unless we were guaranteed security of tenure, in other words political backing. It was a simple formula. Complex base areas attract subversive political elements which in their turn drive out the British interests which the base is there to protect both locally and beyond. In the case of Aden we were providing stability in South Arabia by assisting the Federal Army against dissidents from the Yemen. The base could also support any operations we might undertake in the Persian Gulf. It also gave us aircraft and troops to undertake contingency plans for the emergency evacuation of British nationals in other trouble spots in Africa or the Indian Ocean should the need arise.

But in the end the situation in the base area itself ruled it out as a base in the strategic sense of the word, because most of its numerical strength was required to protect itself; therefore its very existence produced a bigger military problem than the ones

148

it was created to cope with. Thus the strategic potential became completely nullified.

The oft-quoted criticism that after the Second World War Britain lost the will to rule was never more truly demonstrated than in South Arabia. The people in South Arabia were never anti-British, provided we showed that we were strong. Had we banned the political organizations the moment they began murdering defenceless British civilians and loyal Arab police officers, the intimidation of the generally pro-British population and the defection of the Army and Police could have been avoided. No plebiscite was ever held to inquire if the people wanted independence, and in neighbouring French Somaliland 60 per cent of the Arabs voted in favour of French rule continuing. But then the French Foreign Legion did not allow little Arab boys to throw grenades for £5 a time, provided by the Egyptian Intelligence Services. While the British Government made soothing noises to the United Nation's Decolonization Committee the real wishes of the South Arabian people were not even sought.

The decision to develop Aden as a modern military base only made sense if we were prepared to be strong. Firm, tough and timely action to defeat the subversive elements when the first grenade was thrown in December 1963 would have saved many hundreds of lives and stabilized the whole area, including the Persian Gulf. The National Liberation Front (NLF) was a pro-communist party and it would further Russian aims to use Aden as a base because there was no other site in the Indian Ocean which would give her the massive influence she sought.

The Defence Review paid scant attention to long-term strategic interests or indeed the real needs of Britain. The decision to revoke the Conservative Government's pledge to have a defence agreement with South Arabia after independence ensured that a vacuum was left which would only be filled by a hostile power.

This was the 'Little England' policy at its most irresponsible because we wasted 130 years of British rule by leaving nothing but chaos behind. The blame lay at the door of the Labour Government but the Chiefs of Staff share the blame for allowing it to happen and by failing to convince the politicians that the Aden base was always a military nonsense unless harnessed to strong policies.

I had been told in Whitehall by a civil servant of long experience that half the problem for both politicians and administrators was that the Service Chiefs had no collective mind on any subject other than protocol, rank, the honours list and pensions —this was perhaps cynical but a pointer to the difficulties of working a committee system for military leadership.

During the February 1965 visit I was greatly impressed by the High Commissioner, Sir Richard Turnbull, who had recently been recalled from retirement to take over from Sir Kennedy Trevaskis, who had built up great goodwill with the Arab people. But I also heard rumours that Turnbull was not popular with the politicians in Whitehall. I could well believe this as he was a proper colonial administrator of the old school; a man of great intelligence and of physical resilience, one of those men of action and decision upon whose shoulders the British Empire had rested for the past hundred years. I knew this from the splendid reputation he had left behind in the Northern Frontier District of Kenya where for many years he had been the Commissioner. If supported not only by Whitehall but by strong military commanders, Turnbull was a man I felt to be capable of dominating an increasingly dangerous situation and succeeding in maintaining peace in Aden until the British flag was finally hauled down on a new country which had a chance of survival.

But, as I looked round the sprawling base complex, saw the thousands of sun-tanned Service families in their comfortable quarters or on the beach, as I met the Service commanders, all my professional instincts told me that this was wrong. Terrorism —assassination, grenade-throwing and sabotage—was steadily increasing yet most British troops in Aden were deployed more in guarding the safety of their own domesticity than in hunting down terrorists.

When I returned to Aden later with Healey I found the situation had deteriorated further. Terrorism was worse still, the gunmen systematically wiping out the Special Branch officers of the Aden Police, so blinding the Security Forces and making eventual disaster a virtual certainty. Indeed but for the ineptitude and incompetence of the Egyptian-trained terrorists the process would have ended even earlier.

There was no proper state of emergency in Aden as we had known it in other places and, while British social life, within the cantonments, continued as before, the Arab nationalists did much as they pleased in the backstreets. Crater Town was their headquarters and stronghold. Crater *was* Aden to the Arabs.

I had seen terrorism at work before and I considered that the strictest measures necessary to curb it were not being taken. The authorities were ostriches, with their heads so deep in the sand that they could not see that whereas terrorism could be stopped then, in a year or two it would be out of hand. Internal Security operations to combat terrorism are inevitably distasteful but the horror with which critics regard them is partly caused by their own romantic picture of the role of the urban terrorist. In Aden the thugs were hardly worthy of the name terrorist and most of their energies were devoted to intimidating the civilian Arab population. This led to riots, strikes, go-slows, non-co-operation, civil disobedience and general bloody-mindedness. Their role was to bring the Federal Government into disrepute, frustrate progress towards a stable South Arabia, create a void for later foreign influence (Egyptian and Russian), give the impression that the British were being 'driven out' of Aden by superior forces and generally undermine what was left of our reputation in the Persian Gulf. Up-country, where the rural dissident could well be brave and resourceful, it was a different picture, perhaps more reminiscent of the Pathan on the old North-West Frontier of India. Among primitive and volatile peoples, particularly Arabs, a show of strength is not regarded as bullying but is admired; all we needed to do was show our determination.

I was appalled by the complacency of allowing a dangerous situation to become a lethal one but was told of the difficulties. It was difficult, they said, to stop terrorism in Aden because the measures would alienate the civil population and, anyway, there was a shifting population of nearly 100,000 Yemeni workers there, whose movements would be impossible to check or curtail. I thought this to be nonsense. The essence of good internal security is to have an efficient intelligence organization, based on a good police Special Branch, backed by good policemen and soldiers who have confidence in the political handling on the spot. If the home Government is not prepared to back the man on the spot they jeopardize the lives of the civilian community, the police and the army. Squeamish politicians, concerned with their reputations in United Nations circles or emotionally biased organizations like Amnesty International, are liable to take the soft line which leads to increased terrorism and innocent bloodshed. Once the Aden emergency was identified as such it should have been stopped at birth.

The methods needed were well tried and straightforward. In the first place we should have cleared out all Service families and unnecessary administrative units. Then we should have declared a true state of emergency, announced that the penalty for carrying an unauthorized firearm was death and made certain that legislation and action kept in step with such decrees until the end of our rule. The death penalty was invoked in Palestine and Cyprus during the troubles there—had it not been, we could have had a full-scale war on our hands. The movement of Yemenis should have been controlled by the same methods that we used in Kenya to prevent the Somali tribes from moving too far westwards—good local administration. Passes for the itinerant population would not have alienated them if they really wanted work in Aden.

Aden, being almost an island, was very easy to control. The only two overland routes to it are the causeway from Little Aden and the road through Lahej. There is another easily supervised route, along the hard sand of the eastern shore of the isthmus. Check-points could have searched all incoming traffic for arms and the Navy had no difficulty in searching for arms in the cargoes of visiting ships and dhows.

But, without firm military control and with the wiping out of the Aden Police Special Branch officers and, in 1966, the complaints from Amnesty International against the alleged torture of terrorist suspects by the police, meant that the British were helpless in the face of a small but growing force of terrorists, armed and controlled by Egyptian Intelligence Service officers in the Yemen and Cairo. An astonishing procession of idealistic Englishmen—Greenwood, Foot, Shackleton—ensured that our military policies were practically unworkable.

The helplessness of the British was well illustrated in a letter received from an officer in the unit we were taking over from in Aden at the beginning of 1967. He wrote, when we were inquiring about the set-up: 'There is now no deterrent to the terrorist. No terrorist has been executed, although many have been caught red-handed. Captured terrorists are placed in Mansoura Deten-

152

tion Centre where the standard of living is a great deal higher than any of them enjoyed before being caught. The terrorists know they will be released as heroes in a year or 18 months when we go. The only thing a terrorist is afraid of is being caught and shot by an alert soldier or sentry while carrying out his act of terrorism.'

But, in 1965, I was one of the pessimists. Turnbull was rightly pursuing the Government's policy that the Federation could be made to work and affectionately referred to the corrupt, reactionary rulers of the tribal states inland as 'a bunch of bonnet lairds'. In this he was right, and increasingly over the next two years I saw the parallel between South Arabia and the Scotland of two centuries earlier.

I lent people my copy of John Prebble's book *Glencoe* which dealt with the massacre of the MacDonalds, and said: 'Read that and you will understand what is happening up-country. They are a bunch of selfish, inward-looking, feudal chieftains, building up their private armies, pocketing government subsidies, jealous of their richer cousins and sometimes beating them up or having a little massacre, interested not in the future of their country but only in their own little rackets. We should stay on and see fair play because the Arab Socialist powers are linked to Russia, Egypt's long-term aim is to control the Persian Gulf and it is our international duty to ensure security in the area.'

What puzzled me, though, was that a Labour Government in London could support such apparently corrupt and reactionary feudalists.

Denis Healey was an interesting person. He generated confidence. He was erudite, a brilliant lecturer, well-briefed, and he mastered and marshalled his facts with don-like precision. Personally kind and pleasant, he was extraordinarily acceptable in all societies, particularly in the Services, and I believe he was the only Labour Minister the Arab up-country rulers trusted.

Yet in my view he was a shadow without substance. He had all the apparent qualities, the appearance and the methods of a statesman. But I suspected he was entirely dominated by his colleagues in the Socialist Cabinet and their useless dogma, later demonstrated by his series of utterly contradictory defence policies. There must have been a terrible cynicism in him to be party

153

to the Labour Government's fundamental allergy to monarchical institutions in the Middle East whilst having the intelligence to realize that the only long-term alternative was for South Arabia to look towards Egypt and Russia for support.

The most lucid exposition of the intellectual, but to my mind irresponsible, attitude towards South Arabia, I read in the correspondence columns of *The Times* that year. Writing from St. Anthony's College, Oxford, Elizabeth Monroe pointed out that, 'We have to accept that a cold war is in progress in the Middle East, not only between socialist and traditionalist states, but between traditionalist rulers and their own people. In guaranteeing traditionalist states against external aggression, we in effect guarantee their rulers against some of their own people, so involving ourselves in a purely Arab struggle between two groups. Are British interests necessarily bound up with the victory of either? May we not do best if water finds its own level?'

That British armed forces were shortly to withdraw from South Arabia was, of course, known but on the day of the Argylls' arrival in June 1967 we did not know just how soon this was to be. True, the 1965 Defence Review White Paper had announced our withdrawal from Aden by the end of 1968—and in doing so raised the sinking morale of Nasser's forces supporting Yemeni Republicans—because despite the reduction of the Territorial Army, cancellation of the fifth Polaris submarine and TSR2, promises to buy the F111 for the RAF, decisions to abandon the carrier force and virtually destroy the Fleet Air Arm, Defence Secretary Denis Healey had not actually given up any real-estate to please those left-wing socialist circles to whom Britain East of Suez as a responsible power was anathema.

Aden, then, was to be the sop. Combined with the decision to blockade Rhodesia it was a pointer to our pathetic lack of responsible leadership and our desire to be 'Little England'. We were to withdraw not only from Aden itself but also from the wild and backward hinterland which stretched north and east towards the Yemen and Saudi Arabia. This withdrawal, while controlled from Whitehall, had to be carried out by a chain of command established in Aden itself.

The key man in the conduct of affairs was the British High

Commissioner, Sir Humphrey Trevelyan. He had replaced Sir Richard Turnbull only a few months before when Mr. George Brown, then Foreign Secretary, had decided on a change. This was difficult to understand so near to the end of our rule, and particularly as Turnbull had only relieved Sir Kennedy Trevaskis in early 1965. But the interplay of personalities between Ministers and high-ranking officials in the diplomatic and colonial service was a matter outside my knowledge—though one often wondered. In our Regiment we were to get to know Sir Humphrey Trevelyan and come to regard him with the greatest affection and respect. As an individual I shared this view, despite my disagreement with the military handling of events in Aden.

As far as control of the three services was concerned Aden was a galaxy of talent—or, John Masefield's 'bemedalled commanders beloved of the throne, Riding cock horse to parade when the bugles are blown'. Some of these had indeed been riding to parade rather than to the wars in the long period since 1945 and in some cases I was not altogether sure whether they had been sent to Aden because they were best suited for those appointments or because the career structure required them to be given experience to further their advancement. If that was the case they were certainly about to get experience, as indeed we all were.

At the top was the Commander-in-Chief, Middle East Command, Admiral Sir Michael Le Fanu—subsequently the First Sea Lord and long regarded as a naval officer of professional brilliance and personal good humour. His responsibilities were wide and on paper in Whitehall they looked very grand. Although the organization would have been ideal in the old days, when large armies and fleets were spread around the globe and allied strategies demanded large co-ordinating HQs, it was now just the shadow of our former substance and an illusion of power. In fact I was not alone in distrusting the elaborate three-tiered Command structure which made Aden such an unnecessarily complex focal point for the limited British presence in the Indian Ocean and Persian Gulf. Under Admiral Le Fanu the Royal Navy was trying to impose a form of blockade on rebellious Rhodesia and much of his energy had also to be concentrated on the consequences of the war between Israel and the Arabs, which on the day of my arrival in Aden was at its height.

Below the Commander-in-Chief were three entirely separate single service HQs, those of the Flag Officer Middle East, the General Officer Commanding Middle East Land Forces and the Air Officer Commanding. Thus the conduct of day-to-day operations in Aden and its hinterland was the responsibility of his immediate subordinates, men who were to play crucial parts in the events of the next five months.

The General Officer Commanding was Major-General P. T. Tower. He had arrived in Aden only a few weeks before the Argylls, having been the Army's Director of Public Relations at the Ministry of Defence, where I had met him briefly. General Tower was an artillery officer and as far as I was aware he had not taken part in any operations since the end of the war in 1945. This gave him little common ground with the Argylls.

Like the Commander-in-Chief, General Tower worked in the massive headquarters complex on Barrack Hill, Steamer Point, overlooking the Arabian Sea. Its view of blue sea and golden beaches was very different from the squalid alleys and bazaars of Crater, cut off as if in a different world by the thousand foot peaks and ridges dividing them.

Between Tower and myself stood the commander of Aden Brigade, Brigadier Richard Jefferies, an officer with whom I had not soldiered before but who had commanded a Territorial Army Brigade in Ireland. HQ Aden Brigade was originally a static garrison headquarters designed, equipped and staffed to carry out mainly administrative duties. But it was now charged with running a major internal security crisis and it reflected to their credit that until the 20th June they had kept things going, in an area where, despite increasing terrorism, life was supposed to continue with the display of stiff upper lips. No proper fighting emergency, as I would describe it, had been declared and it was only after the incidents which were so tragically and un-necessarily to take place in June that soldiers were officially described as being 'on active service'. This state was achieved after a great deal of personal effort by Jefferies who knew the importance to morale of this particular piece of legislation which politicians so dislike if they are trying to play things down. It might have appeared easy to play it down in Whitehall, but the sad fact remains that during the four years of the Aden emergency the Security Forces lost over 130 killed and nearly a

156

thousand wounded—to say nothing of the European and local national civilians who became casualties.

Aden was, in effect, a peacetime station with more soldiers' wives and children than soldiers. There were thousands of married quarters, ranging from new estates of houses to modern blocks of flats for soldiers' families, all freshly fitted with air-conditioning and refrigerators. There were schools and shopping centres, churches and cinemas. For most of the British, the principal recreation was on the delightful beach clubs, where bars, restaurants and swimming pools had been built under the trees along golden sands beside a sea of darkest blue.

The neighbouring 24th Infantry Brigade, commanded by Brigadier Hew Butler of the Rifle Brigade, was primarily a reserve for contingencies elsewhere in the theatre. It was based on Little Aden, along the coast near the oil refinery. Many of these troops were fighting an old-fashioned 'soldier's war' with dissident tribesmen in the mountains between Aden and the Yemen, a campaign reminiscent of the North-West frontier of India in the 'thirties.

While still in the Ministry of Defence, at the end of 1965, I guessed that the British would abandon Aden in about eighteen months' time. I then heard that among the battalions to be sent there for short tours of up to nine months as part of the force covering the final withdrawal were the 1st Battalion Argyll and Sutherland Highlanders. I knew that the new Commanding Officer, who was to take them there would be myself. This struck me as an excellent arrangement in every sense.

To command one's own Regiment is the high point in a Regimental officer's career, the culmination of years of work and competition. To command it in action, or at least on active service, is a prospect to set the blood racing. I was to take command in January 1967, two months after my forty-first birthday and now, at last, would be able to put into practice all the experience I had gained in twenty-four years' soldiering and world-wide experience of active service.

When I took over command, the battalion was at Seaton

157

Barracks outside Plymouth and I was reminded once again what magnificent soldiers they were—indeed, at that time, the most experienced group of fighting soldiers in the British Army. Not only had many of the officers and non-commissioned officers fought in different campaigns but most of the battalion had only recently returned from the Far East, the only British infantry battalion to complete three operational tours, of six months each, in the Borneo jungle.

There was nothing to change my lifelong view that there was something very special about the Argyll and Sutherland Highlanders. It was not just that Highland Regiments are utterly different from English Regiments because the Argylls were unique in the Highland Brigade too. For the Regiment has not only a record of service and tradition second to none but it has given to the English language a unique expression—'the thin red line'—a sobriquet bestowed by the famous *Times* war correspondent, W. H. Russell, at the battle of Balaclava, when the 93rd Highlanders stopped the Russian Cavalry in full view of three armies and saved the allied cause. (It was not to be the last time that the press were first to appreciate the influence of the Argyll and Sutherland Highlanders.)

This great tradition lived on and I could not have had better material. In the Highlands, the county—or, rather, tribal—regiment still survives and we were essentially a family regiment. Of the thirty-three officers I took with me to Aden, more than half were sons or nephews of former officers of the Regiment. It was the same with non-commissioned officers and soldiers and we had a father and son serving together in Aden—one a sergeant and the other newly joined from the depot.

The Argyll Jock makes a wonderful soldier because of his unique background. They recruit in the industrial belt of central Scotland, westward from the Regimental home in Stirling Castle. The area runs south-west to Paisley and Kilmarnock and north-west into Argyll. They recruit on the coast and the islands to the west of Glasgow and in the suburbs of the city itself. Therefore, the Jocks are urban-countrymen. They have the quick wits of the townsmen in that most live in small industrial towns. They have much of the natural aptitude of the countryman for fieldcraft, because the hills and moors are usually within walking distance of their homes. They are men who have

158

had a tough but homely upbringing. This has tended to make them kind and generous as well as physically hard and resilient.

The Battalion was at that time in Devon as part of 2nd Infantry Brigade. Luckily the Brigadier, James Majury, was an old friend who had been a prisoner of the Chinese in Korea and had then commanded the Royal Irish Fusiliers. We had been at the Staff College together in 1955 and I liked and trusted him. It was typical of him that very soon after I took command he put the Argylls through a testing exercise on Dartmoor with the Guards Parachute Company as our 'enemy' just to make sure that the battalion was as good as we said. The result proved to him that my faith was justified. But I had very strong views about the accepted methods of command and leadership which prevailed in the Regiment, and believed that although the Regiment was vintage brew it always needed a good CO to take the cork out and let it fizz. So in Plymouth I began an intensive training programme, aimed at getting everyone physically fit and mentally adjusted to working to my methods. I was fortunate in having a first-rate set of company commanders, who although at first inclined to military conservatism, made up for this by their enthusiasm.

At the beginning of February I issued a policy directive to trim the ship. It began by stating briefly the cornerstones of my personal philosophy of command. This was: firstly, that everyone can and should enjoy soldiering; secondly, that we were a band of brothers and not a flock of sheep; thirdly, that leadership is a way of life—twenty-four hours a day and seven days a week; fourthly, that physical fitness is paramount; and fifthly, that everyone must know and understand his own job and do it to the hundred per cent satisfaction of himself and the management. I repeated my *aim* to have an efficient and happy fighting battalion with a reputation for being one hundred per cent professional.

It was unique in my experience to have all the rifle company commanders graduates of the Staff College and with successful tours as Brigade Majors or GSO2's behind them. Knowing how Staff College training constipates field soldiers I introduced a deliberately war-time note by saying that: 'Everyone must train himself to act on verbal orders. We establish mutual confidence

159

by cutting out written orders except when they serve a real purpose. For example this Policy Directive has a purpose as it saves me and you hours of briefing incoming officers and NCOs. The CO is the focal point for decision and I am accessible at all times. I also like to deal direct at the level where the decision is needed and not through a time-consuming chain of chaps who should have better things to do. Give me a quick verbal brief and I shall give you a decision. I consider that conferences to collect ideas are the resort of a weak and wet commander. I do business moving around and dealing personally, man to man, with the people who have ideas and problems. I expect people to solve their own problems in the light of verbal policy decisions I give them. When there is a build-up of difficulties I like to have a "think-in" at Company Commander level and highlight the various options. I then decide. I do not believe in personal infallibility and, when I know I am wrong, quickly change my own decisions in favour of better ones. I like to retain a sense of what is possible.' And I added that I liked talking 'shop' and would do so on any occasion, despite the fact that to do this in the Mess or on social occasions was often thought in the Army to be 'bad form'.

I told them, 'In training, think about Wavell's descriptions of the infantry soldier, "quick-footed, quick-minded and light-hearted".

'Leadership. The junior officers must be seen to be leaders, they must mature quickly and be given testing situations which call for a manly response. I do not want any "nannying" of young officers—throw them in at the deep end and tell them you hope they bloody-well drown! . . .'

And, finally, 'We are a fighting Battalion. At all times encourage people to get stuck in. Be offensively-minded and don't pull your punches. Speed everything up and get people cracking. Enjoy soldiering!'

Knowing that Crater was to be our responsibility I wanted to look over the ground and study at first hand the methods of the Northumberland Fusiliers. Later in February, I flew to Aden and stayed a week.

The situation in Aden was now as desperate as was to be expected. Terrorism had been stepped up—both in the number of selective assassinations and grenade attacks. Casualties to

160

European civilians, police and local nationals were increasing. Yet I was told, 'It's interesting to note that anyone can carry a firearm after completing a few formalities to obtain a licence from the police station!' In addition, terrorists were using Belgian-made Blindicide rocket-launchers—anti-tank weapons of about 3·5 in. calibre—as well as mortars of various types and sophisticated booby-traps.

The mounting terrorism was less serious than it might have been because the two main terrorist organizations, both backed by the Egyptian Intelligence Service, had already become rivals for the inheritance of Aden and had begun to shoot at each other as well as at the British. They had penetrated the Aden Police who were therefore virtually ineffective for internal security operations.

Despite all this, the British families were still in Aden, trying to lead normal lives. Their evacuation had at last been authorized, however, and they were all expected to have left by the end of July. By the first of that month, all British troops were expected to have been withdrawn from the hinterland and the rundown to the final evacuation begun. No date had as yet been set for the withdrawal but it was expected to be by the end of the year or early in 1968. At this time there were about 14,500 servicemen in Aden—and about 5,000 in the Persian Gulf.

But my own parish was to be Crater and on Crater I now had to concentrate. The Northumberland Fusiliers had a high reputation and had at that time suffered no fatal casualties. The situation reminded me at once of my experiences in Jerusalem in 1946/47 and in Paphos, in Cyprus, in 1958; it also looked a bit like Zanzibar in '61. In each case, we had found that the only way to dominate a town where terrorists took refuge among the civil population was to move in and live amongst them. And this I planned that we should do.

When I returned to England, the Battalion was being exercised in the Stanford battle-area in Norfolk. I began to explain my methods to my officers, some of whom had had a very conventional training, trying to illustrate, on a small scale the techniques of leadership which had been shown by Moshe Dayan— and which he was soon to demonstrate again—and before him by such men as Rommel and Wingate. Major-General Orde Wingate, who had been killed at the height of his renown in

Burma, was another of my heroes, because of his methods of direct command and personal domination. He had been accused of 'bad form'. But I always remember Churchill's marvellous tribute to him in 1944: 'There was a man of genius who might well have become also a man of destiny.'

That Wingate was one of my few heroes is perhaps an illustration of the bankruptcy of British generalship. The strong thread of brilliant professionalism that ran through German wartime generalship—particularly through its *panzer* leaders—and found its zenith in Field Marshal Rommel, had no echo that was audible to me in the post-war British Army. It lacked 'panache'. It was my view that a self-satisfied approach slowly dominated British military thought after the end of the Second World War to make commanders oblivious to the need for speed, originality, audacity and shock tactics in the new military age we had entered.

Therefore, although the British generals of the 1960s were often well connected, sometimes erudite, invariably charming and of obvious conventional ability they were too often lacking in originality and new ideas. I had little confidence in some of them but there were splendid exceptions to this, men I thoroughly admired and respected .

I began to instil my military philosophy, directly applied to the problems of Aden, as soon as the Battalion returned from Easter leave at the end of March. I was fortunate in that my second-in-command, Major Nigel Crowe, had served for long periods in Aden and had been the GSO2 of the Federal Regular Army of South Arabia. He spoke Arabic, knew the country, had a natural sympathy and liking for Arabs and a host of Arab friends at various levels of influence. Several other officers and NCOs had also served in and around Arabia and I utilized them to the full. We had lectures illustrated by coloured slides and films taken by Nigel. We studied photographs of Crater, talked in Arabic; looked at Arab dress; studied the techniques of the Egyptian Intelligence Service; examined lessons from current operations and generally immersed ourselves in the Aden way of life. I then invited the officer who was then the Army's foremost expert on counter-terrorism to lecture to the entire Battalion under conditions of secrecy.

At this point, I must bring up the question of 'brutality'. Brutal methods are an unavoidable and integral part of terrorism and, sometimes, of counter-terrorism. But I instilled into the Battalion that I would tolerate no brutality and God help the man who was found guilty of it. War itself is brutal and in the half-war of 'Internal Security', roughness that alarms and cows those against whom it is directed very often saves lives. But brutality in the form of beating-up suspects, or torture, I said I would never tolerate and I never did. To make the point clear, on the eve of our departure for Aden I issued a second Policy Directive as follows: 'You all heard the answers I gave to the BBC when questioned about the methods to be employed on Internal Security operations. I sincerely believe that there is no place for thuggery or brutality in a civilized and Christian Army. On the other hand I despise weakness and lack of vigorous and decisive action. The Jock is a disciplined fighter who takes his lead from the top. If the lead recommends brutality he will follow it, and this must be wrong. My policy for Internal Security is best described as 'tolerant toughness'. When the chips are down you must all stick to the rules. I will not condone or allow brutish behaviour. It anyone tries to 'hot' me on this one I shall treat it as a breach of discipline. I have placed Supervision and Self-control at the top of my seven principles for Internal Security operations and I do not want any fancy interpretations of that message.'

I knew that in a few weeks' time we would be flying direct from a soft English spring and early summer into the appalling heat and humidity of Aden and, probably, straight into action.

There seemed little I could do about acclimatization in Devon. But I remembered how Rommel had trained the Africa Corps in the Botanical Gardens in Berlin. So I converted our gym into a steam-heated hot-house and under the merciless eye of QMS I. Wearne of the Army Physical Training Corps we all sweated and sweated until our metabolism was more acclimatized to Aden. Better still, as most of the Battalion had been in Borneo the year before, I could get them in good physical shape by continual tough exercise and training and in this we were lucky in being near the wastes of Dartmoor, where a man's stamina could be tested.

163

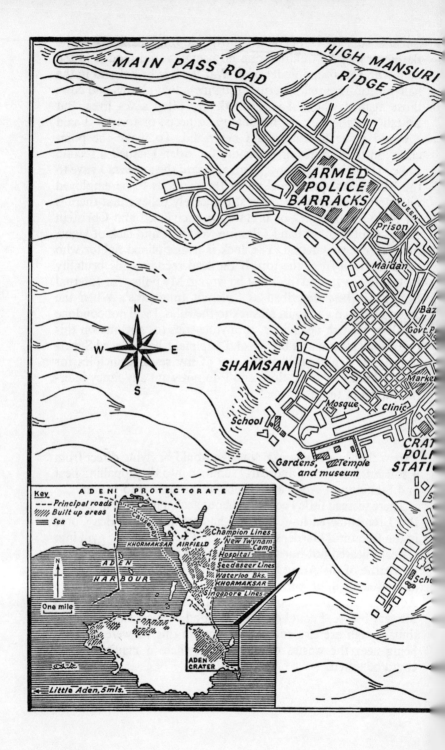

MAIN PASS ROAD

HIGH MANSURI
RIDGE

ARMED
POLICE
BARRACKS

QUEEN.

Prison

Maidan

Baz

Govt. P.

SHAMSAN

Market

Mosque   Clinic

School

CRAT
POLI
STATI

Gardens,   Temple
and museum

N
W   E
S

Sch

Key
– – – Principal roads
///// Built up areas
≡≡≡ Sea

ADENI   PROTECTORATE

Champion Lines
New Twynam
Camp
Hospital
Seedaseer Lines
Waterloo Bks.
KHORMAKSAR
Singapore Lines

Causeway

KHORMAKSAR AIRFIELD

ADEN
HARBOUR

N

One mile

ADEN
CRATER

◄— Little Aden, 5 mls.

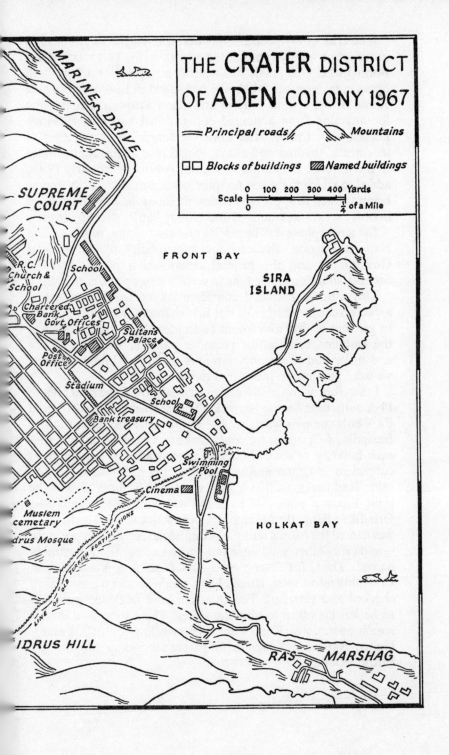

THE **CRATER** DISTRICT
OF **ADEN** COLONY 1967

Principal roads —— Mountains

☐☐ Blocks of buildings ▨ Named buildings

Scale
0   100 200 300 400 Yards
0   ¼ of a Mile

MARINE DRIVE

SUPREME
COURT

R.C.
Church &
School

School

FRONT BAY

SIRA
ISLAND

Chartered
Bank
Govt.Offices

Sultans
Palace

Post
Office

Stadium

School

Bank treasury

Swimming
Pool

Cinema

Muslem
cemetary

drus Mosque

LINE OF OLD TURKISH FORTIFICATIONS

HOLKAT BAY

IDRUS HILL

RAS MARSHAG

The crux of my problem was that the Argylls would have to move into a highly-complicated environment. This was not like Borneo where the jungle was neutral, an equal obstacle to both opponents. Nor was it like any battlefield in Europe, for which Rhine Army, with its armour and heavy weapons, was training for anything from a limited conventional battle to full-scale nuclear war. Crater, 'The Concrete Jungle', a jigsaw town of tenements, shanties and alleys, would give as much advantage to the terrorist as had the Casbah of Algiers in the early 1950s, added to which there was no question of British Security Forces being allowed to apply the ruthless methods the French had used in order to win a prolonged and bloody battle, street by street.

But a new phase was beginning and the collapse of the United Nations Mission, the obvious vulnerability of the Federal Government and the general deterioration of the Internal Security situation had led me to certain inescapable conclusions.

This attitude excluded consideration of purely intellectual aspects, such as whether or not British forces were only in Aden to get shot at. My aim would be to locate and kill terrorists in the battalion area, whilst promoting the maintenance of law and order and protecting ourselves and the friendly elements we were required to guard and escort.

I therefore set up a series of exercises in the Crownhill area of Plymouth, near Seaton Barracks. The married quarters area and the whole complex where we lived were converted into as near a facsimile of Crater as we could make it. One essential was that each building or road should be named after its geographical equivalent in Crater and would be at the right bearing to the other landmarks, so that a Jock would instinctively turn in the right direction when told to head for, say, the Aidrus mosque, Grindlay's Bank, the Maidan or Queen Arwa Road. The NAAFI was named the Aidrus mosque—a strategic feature of importance —and every other building in our barracks complex was suitably named. Then, for forty-eight hours at a time, I would turn Crownhill into Crater. Road blocks would be set up and soldiers checked and searched. The Quartermaster might be grenaded as he left his office to visit the stores. There were cordon and search operations. There were even prowling journalists and a United Nations observer investigating charges of brutality by British troops.

166

The Argylls were also to learn basic Arabic phrases that they would certainly need to know. Two small phrase-books were issued. One was geared to a fairly peaceful existence. It gave the phonetic pronunciation of such exchange of greetings as: 'How are you?' 'Well, praise God.' 'You are *always* well, praise God!' It also gave a brief summary of Arab manners and etiquette. The other was more practical, teaching such phrases as, 'Halt or I fire!' 'Hands up!' 'We are going to search you!' and 'Come with me to the police station.'

By May we were as prepared as we could be in the time available. David Thomson, at this time the Intelligence Officer, and the officers for the Arabic course had already left. I had sent David to Aden ahead of my own party with a special mission. I had great faith in him; he had joined us from Sandhurst in 1963 before we went to Borneo, where he won the Military Cross as a platoon commander. He had been born in Australia and brought up in the Far East and had a wide-ranging and quick intelligence. So, when sending him ahead to learn Arabic, I told him to do more than study the usual Intelligence reports. I told him to study closely the tactical and intelligence techniques and methods while he improved his Arabic fluency. The result was a masterly analysis which confirmed my long-held view that a tougher policy was needed in Aden.

After a delay caused by the sudden war between Israel and the Arabs, the advance party, led by myself, took off from Gatwick in a VC10 on 7th June. This was larger than a customary advance party because of our task. I was accompanied by all my company and platoon commanders, who could then spend about ten days attached to their opposite numbers in the Fusiliers.

By 20th June, the advance party had learned all they could from the Fusiliers and were ready to take over and apply our own methods and the Fusiliers, after nine hard months of tough and dangerous work were looking forward to going home.

Then, just as the first party of the main body of the Battalion was boarding their aircraft at Gatwick, news began to filter out of Crater to the rest of Aden and to the world that a tragedy had taken place. From that moment, noon on 20th June 1967, nothing was to be the same again.

As I waited for the Battalion to arrive I remembered what I had last said to the Argylls before leaving England: 'The team

167

is tough, trained and ready to GO. If anyone by now lacks confidence in my methods they should pursue another profession. The big test is still to come and we shall take Aden in our stride. When the aircraft doors open at Khormaksar the climate will be at its worst and remain so for four months, therefore all ranks must be lean, hard and fit on arrival. The future is full of promise and excitement and all that good professional soldiers could ask for. We must dominate events.'

CHAPTER 11

'Better a foreign grave than native scorn'

GAELIC PROVERB

THE RECAPTURE OF Crater was a wholly successful military operation. Its success depended on a number of factors but the military principle it best illustrated was that of surprise. Somewhere I once read that 'the essence of surprise lies not so much in the nature of an event as in its attendant circumstances. Surprise is occasioned less frequently by an intrinsically remarkable event than by a perfectly normal event occurring under abnormal conditions.'

Before its recapture the situation was as follows. On 20th June all British forces had been withdrawn from Crater. Barriers were established at the two entrances to the town and observation posts, placed on the hills around it, were manned by 45 Commando Royal Marines, initially some of the remaining Northumberland Fusiliers, and the Argylls. These were fired upon from the town and the Armed Police Barracks, where the Armed Police were occupying defensive positions, certain that a British attack would be made on them. The terrorists, quick to take advantage, had established positions in and around the town. A conservative estimate indicated that 400 well-armed men had taken up positions and were prepared to fight. Cairo radio sustained their morale.

The view of the British higher command was that any attempt to retake Crater would have to be made in very considerable force, using heavy weapon support. They feared that this would cause heavy casualties to the Armed Police and possibly to civilians. They continued this gloomy appreciation by assessing that in such an event the South Arabia Army and the Police would mutiny— 'to a man' they said, with that certainty which all good committee men display when they find an excuse to avoid taking positive action. But I had spent long enough in Whitehall to know the stultifying effects of rule by committee. Committees

169

thrive on half measures and compromise, which in Aden lay between the desire to do nothing that could disturb Parliament, the Press and the United Nations and the insistence of good men on the spot wanting to be firm. I had long ago come to the conclusion that the tragedy for the services was that 'Whitehallitis' had spread to many senior military commanders, who took cover behind a smoke screen which they deceived themselves into believing was a sense of responsibility.

They then introduced an even more far-reaching bogey. If casualties occurred the likeliest targets for recrimination would be the 200 or so British servicemen or civilians serving in the Aden Protectorate, mostly in small groups. There was also an RAF radar outpost at Mukeiras which could be threatened and they were convinced that immediate action against Crater would result in chaos throughout South Arabia. I was not convinced by that argument because, as was proved later when NLF overthrew the Federal Government we were pledged to support, the rest of South Arabia did not react to the local Aden situation in any foreseeable manner.

Meanwhile the British civil administration and police were taking measures to reduce the temperature. I did not hear of this in any detail until long afterwards but I gathered they included visits and talks to all South Arabian Army units and action by the Commissioner of Police to persuade the Armed Police to hand back their weapons to the armoury—weapons which had last been used to shoot our soldiers on 20th June and subsequently against our posts around Crater. The Commissioner also arranged for the recovery and identification of the bodies of the twelve British dead in Crater, a macabre and terrible process in which I was personally involved and which probably did more than anything to put the whole matter into perspective for me when assessing the need for firm and timely action as the only possible solution to the Crater crisis.

Arrangements for sealing off Crater included the stoppage of water and electric light supplies, but for 'humanitarian' reasons these were quickly restored and every effort was made to pacify the murderers of our men. In the meantime the Spearhead Battalion of the Strategic Reserve in the United Kingdom was flown out from home on 29th June to reinforce Aden Brigade.

But in spite of all this consideration the conditions in the town continued to deteriorate. Fighting broke out between NLF and FLOSY; hygiene and similar services had broken down, there was a danger of epidemic disease, armed terrorists patrolled the streets and the police made no attempt to interfere with them while Cairo radio accurately announced that a terrorist regime had been set up in the heart of a British colony. Senior officers and administrators relaxed on the cool verandas of Command Hill while only a few miles away, under the Queen's supposed protection, arson, looting and killing went unchecked.

Meanwhile, the measures aimed at reducing tension were continued and the British personnel up-country were moved to positions of greater safety. The usual daily duologue went on between British and Arab officers, each side trying to convince the other that it was *they* who should act to restore order. The Arabs, born intriguers guided by considerations of self-interest, ran circles round their British opposite numbers. They knew that the British were set on ignominious retreat and that the future of the Arab Army was beset with problems of tribalism. But as mercenaries they also knew that as long as the Arab Army stayed in being the stupid British would pay it and, indeed, guarantee its pay and expansion after our withdrawal. If the South Arabian Army re-occupied Crater on behalf of the British—which it could easily do—it would lose its popularity and disintegrate. Better to say to the British, 'It is beyond our capability to re-capture Crater, beyond the capability of the entire Aden Police and the South Arabian Army. Despite the fact that we shall have to rule it eventually *you* will have to capture it.'

The British were stuck, victims of their own attempts at political cleverness. They asked the Arab officers to use their influence to secure a peaceful return, with absolutely no idea what the outcome of such an operation would be. They were like dogs walking round a lamp post, wondering who should put his leg up first. Then, with typical irresolution, they proposed a 'nibbling' operation—a sort of military 'suck it and see'. At that stage I privately determined to take charge in my own way and never look back.

It should be recalled that before 20th June, I had taken a small party of Argylls—about a dozen officers and soldiers in two Land Rovers—into Crater and past the Police station at a time when it was being said that nobody should dare go to this area, but if they did they should go in armoured vehicles. But we went in stripped-down Land Rovers and acted with confidence and without haste. We visited all the supposedly most dangerous areas and we took our time. We walked about and looked around, and although armed and alert we appeared relaxed. Clearly this astonished the Adenis. They had been used to soldiers roaring in and out, riots, ambushes, shooting and tear gas. Now they saw a different sort of British soldier who calmly walked into districts which the terrorists called their own and showed no signs of alarm.

This self-confidence was vital to my method of operating. I remembered how, twenty years before in Palestine, I had been with Cluny Macpherson, my then Commanding Officer, calmly strolling about at a most dangerous moment outside Jaffa exposing himself to enemy fire. But there was method in such madness. Such an ostentatious display of confidence and calm puts as much heart into one's own soldiers as it disturbs the enemy. There is, of course, a risk but it is one worth taking. Once, before 20th June, while visiting Crater I saw one of my officers, who was attached to the Fusiliers, skulking in a doorway. I said, 'What the hell are you doing? Get out into the middle of the street and let people see you!' So once the enemy could see that we were not afraid of them, they became more afraid of us. This was simple psychology and not foolhardiness.

It was not only the Arabs who found us different from any other British troops they had encountered. From the moment our advance party had arrived there was a purposefulness about their behaviour that was so often lacking elsewhere. From the start we stripped down our Land Rovers, mounted our machine guns and *looked* as ready for trouble as we were. We put our Regimental cap badge into our Glengarries, a badge we had lost in some earlier Whitehall futility in favour of a Stag's Head on a St. Andrews Cross, called officially the 'Highland Brigade Badge' but known to the Jocks as 'The Crucified Moose'. It seemed to me that the Army was going haywire, jumping around to suit every politician who had a rush of blood about 'reorganization' and introducing new badges for no good purpose.

As a commanding officer I suppose that I myself was something of a surprise to the Arabs. For one thing, I tried to be frank. I said what I thought. Perhaps the long and to me tedious story of my relations with certain superior officers began very soon after my arrival when I was interviewed on Aden Forces Radio. The interviewer asked me what I thought of the security situation in Aden. I replied that, from the security point of view, Aden was the least buttoned-up place I had ever known. This may have been tactless, but it was true. The interesting thing was that although my superiors may have heard it—or been told by the gossips—none of them took the professional interest to ask me the simple question 'Why?'

My first action after 20th June was to draw up a plan for the re-occupation of Crater. This was explained neatly on a map with all the appropriate boundaries and arrows to show movement coloured in. I put it in my map case and, sure enough, when visiting Aden Brigade HQ, got into conversation with Brigadier Jefferies about a possible re-occupation. 'I've got a plan', I said, and produced it. He registered interest and called the Brigade Major, with maps and a large-scale model of Crater, to join us. This brief initial breath of fire into the bellies of Aden Brigade became a plan which, prior to going home on leave for a rest, Jefferies approved with minor modifications and then got me to present to the GOC and the Brigadier General Staff as 'The Aden Brigade Plan'. In essence it was simple but original. It involved surrounding the north and west sides of Crater from the hill-top positions, helicoptering a force to the Ras Marshag peninsular to come in from the south and simultaneously going in by the Marine Drive entrance, the seaward end, opposite from the Armed Police Barracks, and simply 'rolling it up'.

The actual attack was to be mounted by my own Battalion with armoured car support, while other units 'held the ring' around the rim of the crater and at Main Pass. The attack would take place at night, when most of the enemy would be asleep. And, whereas the enemy did not like fighting in the dark, my Jocks were specially trained in this art.

General Tower, who considered the operation far too much for one battalion, decreed that while we advanced from Marine Drive, 45 Commando Royal Marines would advance into Crater from Main Pass and the High Mansouri Ridge at the opposite end. I thought that not only would this doubling of the assault

173

force be unnecessary but that, as it would be dark, we would almost inevitably end up shooting at each other. This latter argument was unanswerable but it did not influence General Tower who for some reason I could never understand always wanted to talk about infantry tactics as if he understood what it was to be a fighting infantryman. Fortunately the BGS, Brigadier Charles Dunbar, always came in with some intelligent and practical compromise and I personally regarded him as the *de facto* Director of Operations.

Immediately after this conference, Brigadier Jefferies, who was a very tired man and had been ordered to take leave, went home. His place was taken temporarily by the Commanding Officer of the Lancashire Regiment. Before Jefferies flew home, he sent me written orders about the sealing-off of Crater and what action I was allowed to take. In these he wrote, 'You are not to send fighting patrols into Crater until further notice. You may send recce patrols for specific objectives and you are to establish as much observation as possible over the town, with a view to confirming or denying source reports, as well as obtaining an accurate picture of terrorist locations and movements.'

This acknowledgement that I could patrol into Crater itself to gain information was exactly what I had been angling for. It left me with plenty of scope because the definition of difference between 'fighting' and 'reconnaissance patrolling' is academic— a patrol is a patrol. From 25th June onwards I organized an intensive patrolling programme, each in charge of young subalterns full of aggressive spirit. Brian Baty, who as a sergeant in Borneo had won the Military Medal and been commissioned in the field, took the entire Reconnaissance Platoon into the forward edge of Crater in strength, to probe for information. He met no opposition and satisfied me that the terrorists probably slept by night and banged off their weapons at long range by day—a typical Arab approach to the military art.

I was convinced that the intelligence picture painted to us by higher formation was over-pessimistic and probably based on Intelligence Committees taking counsel of their fears. But I could not be certain. Obviously it did not mean that we would not have to fight our way into Crater but our patrol reports led me to believe that there would be no bloodbath on the scale feared and anyway I put more trust in my own assessment of the Crater situation than in that at higher levels.

I was aware of the delicacy of all this but even more convinced of the need for the British to display a bit of grit. But whereas we were full of confidence and winding ourselves up like a spring for the attack, over the mountains at General Headquarters, I suspected that it was still politics before soldiering. When discussing the re-occupation of Crater, they were all for 'playing it cool' and nibbling at Crater, feeling our way forward slowly into enemy territory. I maintained that the only result of this would be to give the enemy generous warning of our approach and enable them to build more defences and set up new firing positions, with the inevitable result that we would be certain of having to fight our way in and my Battalion would suffer unnecessary casualties.

Either the authorities had no idea of the extent of our patrolling or they thought I was exaggerating our competence and professionalism. We had, by this time, reinforced the rocky peninsula of Ras Marshag, at the opposite end of the Crater seashore, sending platoons in by helicopter. But between our forward positions on Marine Drive and Ras Marshag was Sira Island, joined to the mainland by a causeway, and held by the terrorists who regularly fired on us with machine-guns in a routine evening 'hate', but with little effect.

By the morning of 2nd July I was becoming more and more exasperated by these 'wet hen' tactics and decided to do a little bit of daylight reconnaissance myself. With David Thomson manning the machine-gun on my Land Rover and followed by my escort with a second machine-gun I drove through the Marine Drive road block and up to the Supreme Court position where our forward section post was sandbagged in behind the judge's throne.

The Supreme Court road was always under enemy observation and often under fire so it was normal to cover the stretch either in a 'Pig' or else at full speed in a soft skinned vehicle. But as we drew level with the turn in I accelerated rather than braked and headed on for Crater. David Thomson, to whom I had said nothing, cocked his machine-gun and got into a crouched position. My own personal bodyguard, Lance-corporal Hughie Mitchell from Campbeltown in Argyll, was standing

175

upright above my head equally ready for anything. Beside him, Corporals Grant and Logie, my signallers, kept Battalion HQ informed of my movements. We roared straight into hostile territory, past the ice factory to our right and schools to our left towards the burned-out shell of the Legislative Council building, which had been destroyed by mobs on 20th June.

If there were terrorists about they were too startled to fire. But then something alarming did happen. Somebody shouted a warning. I looked round and saw that a low trolley about eight foot long and loaded with Coca-Cola bottles was being pulled into the road behind us, cutting off our escape. Once trapped on the road, we could be shot to pieces as easily as Moncur's party had been less than a fortnight before.

Instant action was essential. I swung the Land Rover round in a tight U-turn and headed full speed for the trailer. We struck it side-on and smashed through, showering Coca-Cola bottles in all directions. The second Land Rover followed, bursting both back tyres on the broken glass. A few moments later we were back at our own positions, elated, relieved but disappointed at having been unable to penetrate more deeply into Crater to get some idea of the layout. The incident was an acid test of my own Land Rover crew and escort on whose quick reactions and coolness my own life depended.

This was the eve of action. If we had achieved nothing else, we had demonstrated once again to the enemy that the soldiers in the red-and-white-diced glengarries were active. With us around anything could happen and probably would. About a month later, an official claim reached the British authorities in Aden for compensation for the breakage of eight hundred bottles of Coca-Cola.

At this time, up to three British Battalions commanded the main road running from Steamer Point to Khormaksar. They were never able to eliminate sniping and grenade-throwing because the back streets behind the flats, office blocks and shops were a haven for terrorists. British patrols, which went into these warrens, were constantly attacked as the time to leave Aden grew nearer. Some areas were considered so dangerous that later on patrols rarely went there at all because to do so would provoke a battle

When we had re-captured Crater I was determined to domin-

176

ate it utterly. Once there we would stay there, we would live there or die there. In Crater, the heart of Aden, the 80,000 citizens would see the Union Jack fly again and understand that this really did represent British rule and law.

On the morning of 3rd July, the acting Brigade Commander came to see me and said that it was acceptable to occupy the small triangle of buildings forward of the Supreme Court, the apex of which was about 400 yards further into Crater than our existing positions and short of where I had had the escapade with the Coca-Cola ambush. I explained that I considered the time was ripe for a deeper penetration, because of the optimistic intelligence picture obtained from our patrolling and that of an SAS (Special Air Service) unit who were operating on the mountain feature and with whom we had an excellent liaison. I believed that this nibbling at Crater would do no more than bring all the opposition down on top of us and we would suffer needless losses. I told him that my patrols had already gone far beyond that point, that we knew that there was, as yet, no serious opposition in the military sense and the only way to take Crater was in sudden, bold bites.

He was entirely sympathetic to the view but had no authority so would put it to the BGS, so I said that I thought we both ought to see the BGS. If there was one thing that always irritated me in the Army it was the needless layers of people passing on orders who had no real responsibility. It is not the fault of individuals but the fault of the system. I personally dislike 'middle men' because they inevitably snarl it up and rarely see the difference between detail and principle. In Aden, where everyone breathed down the neck of everyone else in the chain of command, it was ludicrous.

So we went to HQ Middle East, where I saw Charles Dunbar. I stressed that I would like to go deeper into Crater and seize it by the throat. Dunbar, a proper soldier, saw my point and authorized me to go up to the limit of my first Phase line but not beyond under any circumstances without permission. This was great news as the first phase was a really large slice of real-estate—a good quarter of the Crater District.

I left Dunbar, fully satisfied, but I was told months later that General Tower was lunching at Little Aden and confided that that night we were planning 'a little probe into Crater'. Just what

177

this 'little probe' was to be and how great its world-wide repercussions could not have been apparent to him until the next day. Perhaps this was the root cause of our future disagreement.

I went back to Waterloo Lines and held an Orders Group for the Company Commanders. I did the briefing on a large model of Crater which showed the minute detail, the relative position of streets and houses and the height of buildings. As they were leaving I asked Major Ian Mackay, who had been the Battalion part-time Public Relations officer before he took over 'D' Company on Bryan Malcolm's death, if he had arranged for any press coverage. He said 'not yet' so I told him to fix it up. I had always believed that it was best to brief correspondents directly yourself, and in the Ministry of Defence, in Mountbatten's day but not afterwards, I had gained some experience at dealing with the top flight of defence correspondents.

Later Stanley Bonnett of the Associated Press and the *Daily Telegraph*, Stephen Harper and the brilliant war photographer Terry Fincher of the *Daily Express*, Anthony Carthew of the *Daily Mail*, Barry Stanley of the *Daily Mirror* and John Dodd of the *Sun*, arrived at Marine Drive, where most of the battalion was waiting ready in a large quarry beside the road and out of sight of the enemy. I told them, 'We're going into Crater'. No one came to watch the start of the operation and no Army Public Relations Officer came to shepherd the press.

The moment for action drew nearer. That afternoon, I had sent two more platoons by helicopter to the Ras Marshag peninsula, so that when the attack began, we could start to 'roll up' Crater from two start-lines at the seaward end of the town. Both advances were to be led by Major Paddy Palmer's 'B' Company, that from Marine Drive being supported by Saladin armoured cars under Major Tony Shewan of 'A' Squadron Queen's Dragoon Guards. Then as the main force followed up, Major Ian Robertson's 'A' Company would push out along the causeway and take Sira Island.

The force advancing from Ras Marshag would, initially, only move up to the edge of the town, concentrating on occupying the high ground above it and the old Turkish fortifications from which any opposition to the main advance from Marine Drive could be out-flanked. But another vital part of the plan was for Lieutenant Hamish Clark's platoon, which had for several days

178

been infiltrating on to the top of Aidrus Hill (a dominating feature more than 700 feet above the South West corner of the town), to supply observation and fire support if anything went wrong. We expected to complete the first Phase and have advanced on a continuous front from the sea to form a line some 400 yards inland, so occupying about a quarter of Crater, by 10 p.m. If all went according to plan, we would then advance in two more Phases, immediately afterwards into the heart of Crater, occupying more than half of the town.

By seven, it was almost dark and the unsuspecting terrorists on Sira Island were getting ready to start the inevitable shooting and blindicide firing which punctuated the early evening. The Argylls were ready at the start line, awaiting my order to advance.

As we waited on the start-line *nobody on the British side knew how much opposition we would have to face.* Remembering the killing of our friends on 20th June, many were perhaps hoping for strong opposition so that they could avenge their murders and defeat the enemy once and for all. But I was determined to avoid any recriminations of that nature. None of us needed reminding about these lost friends, but Ian Mackay carried into action the cromach—shepherd's crook—that had belonged to Bryan Malcolm and every armoured car of the Queen's Dragoon Guards had, tied to the top of its wireless aerial, the red and white feather hackle of the Royal Northumberland Fusiliers.

It used to be traditional in Highland Regiments to be piped into battle and the custom survives if and when the tactical situation allows. The younger generation of officers and soldiers had never seen it happen, so, when training at Stanford the previous February, I had purposely staged a dawn attack with the Pipe Major playing along the axis of advance beside me. It is the most thrilling sound in the world to go into action with the pipes playing, it stirs the blood, reminds one of the great heritage of Scotland and the Regiment. Best of all, it frightens the enemy to death! In an Internal Security operation against at lot of third-rate, fly-blown terrorists and mutineers in Crater on 3rd July 1967, it seemed utterly appropriate. I ordered Pipe Major Kenneth Robson to sound the Regimental Charge—'Monymusk'. As he began the Jocks started to move down the road leading from the start-line into Crater.

179

In addition to my own Battalion I had under command 'A' Squadron of the Queen's Dragoon Guards in their armoured cars; a troop of 60th Squadron Royal Engineers; a helicopter from 47th Light Regiment Royal Artillery; a rear link wireless set from 15th Signal Regiment Royal Corps of Signals and additional transport from 60th Squadron Royal Corps of Transport. All of these soldiers did magnificent work; it is this co-operation of various types of unit which brings success.

Hardly had we started than we were machine-gunned from the edge of the town. Everyone bit the dust—with a few notable exceptions! The Pipe Major oblivious to the noise of shooting, played and marched on. Our forward section in the Supreme Court began to return the fire. I climbed up on the side of the forward armoured car and spoke to the commander to see if he could identify where the fire was coming from.

Paddy Palmer, a few yards ahead, came up on the wireless saying, 'They're firing from either side of the Sultan of Lahej's Palace'.

I told the armoured car commander to brass them up and ordered the advance to continue. Despite the weight of fire we put back at them and although the terrorist machine-gunners eventually stopped, a single sniper on the roof of the Sultan's Palace continued to fire. His bullets pinged above our heads, well above I thought, so I ordered no retaliation as it was too dark to return accurate fire at that range. I walked back to my Land Rover a few yards away, took up the radio microphone and said to all stations, 'Play it cool'. I was quite determined that our fire control should be absolute until we met the main enemy positions which I believed to be in the centre of Crater.

Captain Robin Buchanan and the rest of 'B' Company, advancing to join up with us from Ras Marshag, had had to kill the only man to die that night. Near a cinema on the outskirts of Crater, they called on a group of Arab men to halt. They did so, but one armed man made a dash to escape. He was shot dead. That, indeed, was the extent of the 'bloodbath' so gloomily forecast.

'B' Company carried on and I walked along beside my Land Rover. The newspaper correspondents, and Terry Fincher came too. They were utterly co-operative and friendly and shared the experience in every sense. This is why the Press were more

180

reliable guides to the activities of the Argylls in Crater than many people in authority.

Within an hour we had established an observation post on the Chartered Bank and in the shell of the burnt out Legislative Council Building. I sent Ian Robertson across our rear with a mixed force of armoured cars and infantry to take Sira Island. He dashed off like an express train. The Jocks climbed its steep sides in the dark and clambered up over the walls of the old Turkish forts. The terrorists had flown, leaving their banners and flags limply hanging in the hot darkness of a South Arabian night. These were hauled down from the topmost flagpole and an Argyll flag put up.

My next problem was to take over the Treasury Building. This contained the whole of the treasury reserve currency for South Arabia and was occupied by the Armed Police. There was no way of knowing how they would react to our appearance so I decided to send Nigel Crowe with the assault platoon to see if he could charm them into submission. It was a dramatic performance. In the circumstances we would have been justified in shooting it out had the Armed Police sentries offered any resistance. But I continually stressed the need to avoid unncessary bloodshed and, rather whimsically at that moment of danger and tension, my mind went back to my schooldays and to Plutarch's account of the life of Marcus Cato: 'Accustomed as he was to hard exercise, temperate living and frequent campaigning so that his body was healthy and strong, he also practised the power of speech, thinking it a necessary instrument for a man who does not intend to live an obscure and inactive life. In battle he was prompt, steadfast and undismayed and was wont to address the enemy with threats and rough language, rightly pointing out that this often cows their spirit as effectively as blows.'

Nigel, with his usual courage, stood out in the open street and negotiated in Arabic, pointing out that we were not going to kill them but intended to occupy the building. It was a tense moment but he gradually won their confidence, they opened the steel doors of the Treasury and, although showing signs of extreme nervousness, accepted our occupation of the building.

It was a brilliant bit of work by Nigel and the NCOs and Jocks who were with him. While it was going on Ian Mackay was exploiting quickly and with a feeling of exhilaration I realized that we were well into the second phase of my original plan and it was not yet 11 o'clock—four hours from the start of the operation.

It was now obvious that we were over-reaching the limits of exploitation agreed by Charles Dunbar, so I spoke to the acting Brigade Commander on the wireless and said that the initiative was so completely ours that to pause might lose it and he should ask permission to let us go on and exploit up to the civil Police Station, where I planned to repeat Nigel Crowe's successful tactics at the Treasury. The one thing I was determined to do was retain the initiative and not stop.

This was the rewarding moment for any commander, when you know that your own chaps have got their tails up and will cut through opposition like a knife through butter. You can feel it in the air, and breathe in the aggressive confidence. It is the battle-winning factor that only experience can gauge. To me, that single moment in Crater was worth all my quarter of a century of soldiering. I felt, as we all did, thrilled to be an Argyll and to be writing another chapter of Regimental history in the tradition of our forebears.

We were in luck. Charles Dunbar gave us the 'green light' to go ahead as far as the Civil Police Station. I ordered Ian Robertson to take another mixed infantry and armoured car group to escort Nigel's Arabic speaking party, and off they set. It went without a hitch except that during the complicated re-grouping of his company and the armoured cars Ian Robertson and his Company Headquarters were fired on from an armoured car which mistook them for a group of terrorists. It was typical of the tremendous spirit of co-operation which existed between ourselves and 'A' Squadron of the Queen's Dragoon Guards that the situation was very quickly restored and the regrouping of the Jocks and the cars completed, without imposing any delay or halting operations because of the risk of further mistakes of identity, which in the circumstances were almost inevitable. This incident confirmed my earlier contention that the re-occupation was best carried out by a single battalion because of the difficulties of controlling fire in the dark.

By the early hours of 4th July I was fully confident that our aggressive and spirited behaviour had frightened the life out of any potential enemy. What was more important, the feel of the

Battalion was good and I knew we were capable of exploiting our initiative to the full.

Opposition had been slight, but we had to remember that the most dangerous areas—those surrounding the Armed Police Barracks and the Aidrus mosque, which was a terrorist stronghold—were still an unknown quantity.

But we were well satisfied with the night's work. Our 'little probe' had given us half of the enemy's territory and he now knew that he was up against British soldiers who had come to stay.

Some of them must have recognized that we wore on our heads the same red and white glengarries as had three of the British soldiers they had seen so treacherously killed on 20th June. I expected that they were frightened. I hoped that they were.

The snap and whine of bullets and the armoured cars prowling through the narrow streets had kept the citizens of Crater behind locked doors. When dawn broke on 4th July they heard a new sound, one that was to remind them that, until the British finally left South Arabia, here in Crater the rule of law would be enforced. They heard the Pipes and Drums. On the roof of the Educational Institution, overlooking the flat roofs and minarets towards the horseshoe of mountains and the Arabian Sea, our pipers played. We had riflemen guarding them but they played as well and with as much composure as if they had been back at Stirling Castle.

It so happened that just before we had crossed the start-line at 7 o'clock someone asked me what was the code-name for the operation. I realized that on the senior levels where such matters are usually documented with care, the whole business had been seen so vaguely that it had not been given one. I therefore then and there decided to call it 'Operation Stirling Castle', after our Scottish home.

Early on the morning after the attack, visitors began to arrive. First, television crews—both BBC and ITN—and correspondents who had either been away from Aden or had just arrived an hour or so earlier on a flight from London.

World reaction had been immediate and intense. Later we were accorded a visit from General Tower. Although tired, I was relaxed and happy and my welcome was friendly. I sensed

183

that he was hostile. 'Everything is going very well' I assured him. 'Would you like to see the map?' He was frigid, his voice carrying a note of veiled sarcasm which I was to hear so often in the months ahead as we fought to dominate Crater. After hearing my brief report, he said, 'This is just what I had planned.'

From that moment the GOC hardly spoke again on that morning. I took him up the road in my Land Rover and he seemed ill at ease, asking, 'Where are we now?' I thought that he should visit the Police Station and see the extent of our grip on Crater.

When he had gone I spoke to Nigel Crowe but said nothing about my personal feelings. It is an obvious requirement of military loyalty that you keep your own counsel if you have no faith in your superiors or else ask to be relieved of your command. But in the Argylls we had all soldiered together for a lifetime. Nigel, whom I had known since he joined us in Palestine twenty years before, sensed my mood. Sleep and rest are two of the most important factors in sustaining morale and Nigel had caught up on a bit of well-earned sleep that morning to be fresh to relieve me to do the same. So after Tower's visit I told Nigel that I was going off to sleep for three hours and in the meantime he should continue the consolidation and 'make contact with the Armed Police'.

When I returned at 5 o'clock Nigel told me that he had, in fact, managed to telephone Superintendent Ibrahim at the Armed Police Barracks, where the mutineers were, understandably, getting extremely agitated at the thought of retribution which they believed imminent.

Now Nigel knew—and liked—Arabs and he realized that this was no moment for soft diplomacy. The potential enemy, still powerfully armed and at least equal to us in numbers was still in position in their half of the town. So, on the telephone, he ordered Ibrahim firstly to guarantee that, when we moved up Queen Arwa Road to Main Pass to occupy the whole of Crater, there would be no opposition from any quarter, particularly from the terrorists who had fired from the flats across the road from his barracks on the 20th June. Unless such a guarantee was forthcoming, Nigel gave him to understand the Armed Police barracks would be utterly destroyed. Then he ordered Ibrahim to hand over all the mutineers of 20th June, so that justice could

ARGYLL JOCKS

RGYLLS IN THE
ORNEO JUNGLE
1964

SUE WITH COLINA, ANGUS AND LORNE, 1967, WHILE THE
AUTHOR WAS IN ADEN

AIR VIEW OF CRATER. SEE MAP, PAGES 164-5

LITTLE ADEN JUNE, 1967

PIPES AND DRUMS PLAYING REVEILLE ON CRATER ROOFTOP, THE FIR
MORNING OF REOCCUPATION

Photo: Daily Express, Terry Fincher

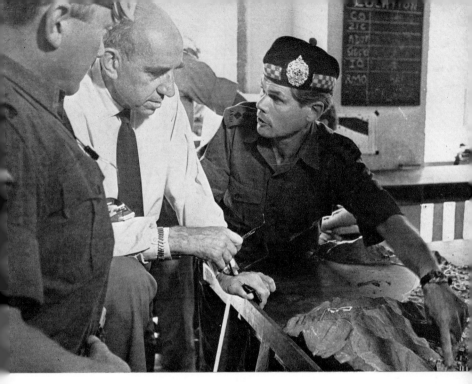

THE AUTHOR BRIEFING SIR HUMPHREY TREVELYAN, THE HIGH
MMISSIONER, AFTER THE REOCCUPATION. NIGEL CROWE ON THE LEFT

Both Photos: Daily Express, Terry Fincher

TH THE SUPERINTENDENT, ADEN ARMED POLICE, THE SECOND DAY
OF THE REOCCUPATION

AN ARGYLL OBSERVATION POST

THE AUTHOR (BACK TO CAMER.
TALKING TO IAN MACKAY
AND DAVID THOMSON.
ON LEFT, MAJ-GEN. TOWER

Photo: Whyler, Stirling

ES AND DRUMS OF THE 1ST BATTALION LEAVING STIRLING CASTLE, LED BY CRUACHAN, THE REGIMENTAL MASCOT, MAY 1968

E AUTHOR, AT A TRAINING CAMP IN NORFOLK, ANNOUNCING THE GOVERNMENT'S DECISION TO DISBAND THE ARGYLLS

Photo: Daily Express, Terry Fincher

AFTERMATH. AS A CIVILIAN WAR CORRESPONDENT, VIETNAM,
OCTOBER 1968

take its course. He assumed, as we all did, that the guilty mutineers would be brought to trial and punished. They were murderers.

Ibrahim pleaded that he could not possibly guarantee that there would be no opposition. But he said he would try to identify the mutineers and then hand them over—something which for reasons I have never understood was never allowed to happen. I informed the acting Brigade Commander of the contact and suggested an early occupation of the rest of Crater. I was given permission to move nearer the Armed Police barracks and at 6.30 p.m. sent 'A' Company up to the line of the Haddadin Bazaar with orders to patrol extensively to the south west but no nearer to the Armed Police barracks. This operation went off without a hitch and set the scene for the final act.

Ibrahim had been shaken. I invited him to meet me outside the Chartered Bank, early next morning, where I privately gave him to understand that if, when we advanced again, there was any trouble, I was prepared to wipe out the Armed Police to a man. I told him that my soldiers were fresh from the Indonesian War and that we were 'hillmen' like many of his armed policemen, who believed in deeds rather than words. He got the point!

But by now the wires were humming and the full machinery of administration and government were analysing the effect of the Crater re-occupation. The first man to arrive to see us was the High Commissioner, Sir Humphrey Trevelyan. He was a man I always enjoyed meeting but on this particular morning he did great things for our morale. He was enthusiastic, practical and utterly human. I warmed towards him as he said how well the Jocks had done and what a splendid effort the whole thing was. I was his servant from that moment onwards and throughout the months ahead, though in my inner and private thoughts I thoroughly disagreed with the dishonourable policy of the British Government which he implemented so brilliantly. The feelings of the Argylls towards him were affection and respect.

At 11 o'clock I saw Ibrahim again and we had a meeting at my Headquarters with Peter Owen, the Commissioner of Police, and his Deputy, Said Abdul Hardi Shihab, together with Brigadier Dunbar and, on and off, General Tower. The meeting decided on policy for the future and it was agreed that the Armed

185

Police were to be left in Crater so that we and the civil police could all 'co-operate' together.

It was made obvious to me that such a charade was going to be the official policy and I agreed that we should do our best to make it work. I had been in some strange situations in the old Colonial Empire but this was surely the most grotesque. Owen, the Commissioner of Police, had the impossible task of policing a colony where official policy was now condoning the murder of British soldiers by armed policemen in the interests of a political settlement. His Arab Deputy was about to inherit a subverted, politically split and badly frightened organization which had been a popular target for assassination. He had seen too many loyal Arab policemen gunned down to have any faith in the British and yet he was ambitious for control after Owen's departure. Mohammed Ibrahim had acted as a sort of dishonest broker, trading British corpses to try and preserve his own skin. This was all going on in a colony which until recently British officials had described as 'vital and necessary to the defence of the free world!' With typical self-delusion, or perhaps it was just old English humbug, the Government continued to cherish the illusion that it could not only withdraw from Aden but at the same time establish a pro-British regime to leave behind it. As the meeting ended my feelings were mixed but I was more and more convinced that the safety of the Argylls depended on our own Regimental spirit—for we were indeed contemplating strange bedfellows in the next act of the drama.

Ibrahim told that when we completed our advance he did not expect opposition. At five o'clock that day, the road block at Main Pass was ordered to be opened, and no doubt some of the more guilty were allowed to slip away in the process.

At 4 p.m., a column of the Argylls and the Queen's Dragoon Guards roared up Queen Arwa Road to the two charred wrecks of Moncur's and Malcolm's Land Rovers. There the column halted and while Jocks took up firing positions, the wrecks were examined for boobytraps and then prepared for towing away from the scene of tragedy. I swung off the road and drove into the Armed Police Barracks where I was saluted by Ibrahim and saw no signs of resistance. My thoughts were very mixed as I was only a few yards from the point of Bryan Malcolm's death and I thought of all I had seen from the helicopter on the 20th

186

June. But now it was as good as over. The column started up again and, with the first of the burned-out Land Rovers in tow, drove up the road to Main Pass to complete the occupation of Crater. I saw the Acting Brigade Commander drive in from the Main Pass end with an escort from the Parachute Regiment.

Throughout the two-day operation we had not lost one man killed or wounded.

The general feeling of relief at home was perhaps best expressed to us in a leading article in the *Daily Telegraph*, which said, 'British troops have shown the combination of skill, tact and cool courage for which they are unequalled by regaining control of the Crater district of Aden. . . . Thus a dangerous and humiliating state of drift, which the British Government permitted to continue for a fortnight, has been ended. The longer this lasted, the more credible did the terrorist boast become that Britain would likewise be driven out of the rest of Aden long before independence. . . .'

The dramatic story of the re-capture of Crater had ended. But for us, the Crater story was only beginning. Now we were going to live there.

CHAPTER 12

*'The apprehension of personal danger can easily be mastered once
the lesson has been learnt that nothing worse than death can be
expected.'*

VLADIMIR PENIAKOFF ('Popski')

IT WAS NEARLY thirty-six hours since we had started the re-
occupation. I needed a new Battalion Headquarters but was
uncertain where to go. Siting an HQ is a matter of compromise
between the conflicting demands of wireless communications,
security and administrative convenience. Alastair Howman, who
commanded HQ Company, and Hugh Clark, the battalion signals officer, put forward all the pros and cons. We decided on
the Chartered Bank Building. Not only was it the best constructed and most modern building in Crater but it also dominated the whole of Queen Arwa Road and therefore was a key
spot in controlling the two entrances into the town. Another
factor in its favour was its accessibility. It was easy for visitors
to get to either from Marine Road, or Main Pass.

But above all it was the sort of building from which we could
fight and control the companies should the South Arabian
situation deteriorate into street fighting. At the back of my mind
I was always prepared for the worst case—which seemed to me
to be an attack by the South Arabian Army using heavy
weapons.

Some people criticized my choice and told me I was foolhardy.
The Bank would become an obvious terrorist target. Indeed,
judging by the bullet holes in the windows and the marks where
blindicide rockets had smashed into the building, this was stating the obvious. But those who held this view did not understand either me or my methods. If we were going to continue to
dominate Crater, in the way I planned and for the weeks and
months ahead, it could only be by active leadership and a tremendous effort on the part of everyone in the battalion team.
Above all, there must be no question of a split occurring in the

Battalion with the rifle companies constantly exposed to danger whilst Battalion Headquarters occupied a comfortable and safer billet elsewhere. The whole of my philosophy of command rested on identification with the Jocks in the rifle sections. If anyone was going to deploy soldiers to live permanently in Crater he must obviously do the same himself. By the end of our occupation in November I was the only officer or man in the Battalion who had not slept out of Crater one single night throughout the period—that was how it was to be and how it should be.

Thus Battalion Headquarters moved into the Chartered Bank and, as my critics predicted, and I will describe later, it did become a prestige target.

The usual bank entrances on the ground floor were sealed off and we kept open a single side entrance. The Manager's office became the Regimental Medical Aid Post and the remainder of the open-plan floor space and counters became the Jocks' Dining Hall and the Interrogation Centre. The nerve centre of Battalion Headquarters, the Operations Room, was in the Rolex Watch Company offices on the first floor. Throughout the five months that followed we remained there and so did all the watches.

The second and third floors were devoted to accommodation for the officers and Jocks. At the top of the building were two penthouse flats which had been occupied by the British bank staff. One of these was bare though it eventually doubled as the Battalion Kirk and Sergeants' Mess; but the other was well furnished down to reproduction French Impressionist paintings and expensive Turkoman rugs. This became the Mess where the eight officers of Battalion Headquarters and I lived for the next five months.

On the roof of the two flats we placed two look-out sentries, usually manned by the Pipes and Drums. These two posts played an important part in the domination of the town.

Now at last my methods of controlling Crater were given a chance to be put to the acid test. This was the logical development of my experience in Jerusalem and Cyprus and of what I had seen in Zanzibar. It was very different from the Aden Brigade method. First of all I divided the town into three parts, one for each rifle company. As opposed to normal Internal Security practice, I did not make these coincide with the Police Boundaries as I did not want the Police to understand our layout. I

189

did not trust them. 'A' Company were responsible for the northern end including the road block on the Main Pass, the Armed Police barracks and the Maidan where most of the best shops were located. 'B' Company had the coastal area and the Marine Drive road-block. This area was more open than the other two, having many old buildings dating from the days when the British Garrison had been billeted there. Apart from the now burnt-out Legislative Council Building, which had once been the Garrison Church, there was the Sultan of Lahej's Palace, an old barrack building. There were also a number of smart blocks of flats and well-appointed houses some of which had, until quite recently, been occupied by British Service families.

I gave 'D' Company the centre of the town, where the streets were narrowest and the population most dense. The Market Place was here and so was the Civil Police Station. The Market Place had always been a notorious centre of terrorist activity and in the weeks that followed it certainly lived up to its reputation.

Lastly, there was Headquarters Company. It had a number of independent platoons: the Reconnaissance Platoon occupied the Observation Posts on Temple Cliffs, the old fortifications on Mansuri and the Brown's House Ridge which divided the town from the Ras Marshag Peninsula. The Assault Pioneer Platoon, who later on reinforced the Reconnaissance Platoon, were employed flat out on building defensive works and sandbagging Observation Posts. The first of these was the new road block on the edge of the town itself. This was completed on the night we went in.

The Pipes and Drums were the Battalion Headquarters' defence and vehicle escorts. The Military Band, who are a Silver Band as opposed to the Pipes and Drums and who are so often regarded as musicians in uniform and not proper soldiers at all, had the misfortune to have their newly ordered set of instruments stuck the far side of the Suez Canal as a result of the Arab-Israeli War. They buckled to and made an invaluable contribution helping on the road blocks and countless other tasks.

The secret of our controlling Crater lay in the fact that virtually the whole Battalion lived there night and day. On average thirty different posts were established at any one time throughout the town. The tops of high buildings in commanding posi-

190

tions were fortified with concrete blocks and sandbags so that most of the main streets were covered by Observation Posts and interlocking arcs of fire from machine-gun positions.

Another vital part of the method was that every post, patrol and vehicle was always linked onto the Battalion Radio Net. Every Jock in the Battalion, not just the signallers, became used to speaking on the radio, reporting incidents or relaying routine reports. This had tremendous advantage because it meant that any incident or report of suspicious activity anywhere in the battalion area was immediately known by everyone. This gave the isolated posts confidence. The instant passing of information more than once resulted in the killing of a grenadier who otherwise might have escaped unseen into a back alley. It also brought some extraordinary harassing hours for Hugh Clark, who as signals officer, had to control the 'net'.

Tactically, I was determined to keep the enemy guessing. For instance I might order one day that every Arab riding a Honda motor-cycle should be stopped and searched thus giving the impression that we knew something about an Arab on a motor-cycle of this make (which we probably didn't). Another day the focus might be on Opel Taxis. We patrolled continuously, supported by the menacing armoured cars, which had a strong psychological effect on potential trouble makers. The patrols were varied and irregular so as to cause maximum confusion.

Our joint patrols with the Queen's Dragoon Guards, later replaced by the Queen's Own Hussars, were most effective by day. We found that at night our foot patrols were the best answer as the terrorists were afraid of the dark, and night attacks were very rare. At first, all street lights were turned off. But, after a few days, orders came to the civil authorities that lights must be switched on again. This meant that our night patrols became intensely vulnerable to sniping or grenading by terrorists lurking in dark alleys. So we put most of the lights out again for the sake of our own safety.

Another 'ploy' was the 'phantom' observation patrol. We would occupy a flat secretly and at night. In the morning, taking care not to be seen, which can easily be achieved by silence and keeping back from windows, the patrol would have a fresh view of the town. The terrorists took a lot of trouble to find out which buildings we occupied so that, when violence was contemplated, they had a reasonably safe get-away. But more than once this type of incident was nipped in the bud because we kept them guessing.

Now that the re-occupation was complete and the battalion was deployed throughout Crater the real task for which we had been brought out from the UK could begin. We had entered Crater by surprise. Now we had to hold it by strength, skill and, above all, courage.

To those who lived within the rim of the encircling hills, our methods during the first days of July were obviously a new manifestation of British colonial rule. I have described the attitude of our predecessors which was based on the official premise that the presence of British troops encouraged violence. We had now to prove that the exact opposite worked better. And we had to prove it not only to our own side but far more importantly to the citizens of the 'kingdom' we now controlled.

I had instilled into my Jocks the knowledge that they were superior to any enemy they might meet. Even the possibility of a mutinous South Arabian Army posed a very hollow threat to men who had spent much of the previous two and a half years facing the Indonesians in Borneo. Some battalions were being trained to be cautious and beware of the terrorist sniper or grenadier. I wanted the Jocks to hunt and kill terrorists: they had entered Crater without fear and this initial success had added to their self-confidence. This confidence of ours obviously lowered the confidence of our enemies; the initiative had been wrested from them and it was up to me to see that they never regained it.

On 5th July, when we commanded the whole of Crater, I had hoped that we would be allowed to keep it sealed off with road-blocks on Main Pass and Marine Drive to give us some days during which we could find our feet with the inhabitants. For some reason which I never understood and which we all greatly resented at the time, we were not allowed to do so, and on that day Crater had to be open to traffic. We could search civilian vehicles, but not those of the Aden Police, the Armed Police and the South Arabian Army, which meant, of course, that enemy

192

arms, reinforcements and agents could move freely under this cover.

My first act was to assure the Battalion that whatever the orders or whatever the consequences, no Argyll would ever be abandoned in Crater. I told the company commanders and the RSM that if any repetition of the 20th June incidents took place they were to counter-attack without reference to me, or anyone else, and that this was the order all the way down the line. There would be no excuses. I said, '*If you have no ammunition you are to go in with the bayonet. It is better the whole Battalion dies in Crater to rescue one Jock than any one of us come out alive.*'

This message was quickly passed and I know that it did much to steady the nerves of some of the younger soldiers who were on active service for the first time. Nor was it bombast, for this was the basis of my confidence in leaving small groups of Jocks isolated throughout Crater—they knew that if anything went wrong the Battalion orders were for instant rescue action with no holds barred.

We set about dominating by strength of character and showmanship. Already the Adenis realized that we were different from any other British troops they had seen in our methods and behaviour; my rumour soon spread that we were wild Bedouin tribesmen from the Scottish Highlands who could get very rough indeed if provoked. The officers in the Argylls always wear grey shirts in the East, a tradition dating back to pre-war service on the North West Frontier of India; therefore I was amused to hear a report from the editor of a local Arab newspaper that according to rumour in the bazaar, 'You are a Jewish penal battalion and the ones in the grey shirts are the Rabbis!'

I had long suspected that the Aden taxi drivers were a source of terrorist information and a means of transport for gunmen and grenadiers. We went out of our way to curb the speeding and hooting taxis from dominating the streets. They soon realized that the speed limit applied and horns were taboo, for we knew that they could be used as terrorist signals. We also changed the traffic rules and made a one-way circuit for all but our own vehicles. Our own vehicles chose whichever way the driver cared to take. In the narrow bazaar streets it was all too easy for military vehicles to be blocked and stopped, apparently by accident. I had personal experience of this with the incident of the Coca-Cola lorry. Once stationary, they provided a sitting target and were incapable of moving to assist if an incident occurred nearby. We demanded, and achieved, the same freedom of the road that a fire engine would expect in this country.

There was no official curfew in Crater but it became obvious that we suspected anybody on the streets after dark and I told the civil police that for everyone's safety it was best if they advised the citizens to go to bed at seven o'clock. The night provided a cloak for the experienced terrorist and if he has a crowded Arab street to disappear into his task is all too easy. By clearing the streets at dusk we removed much of their cover and the lack of street lights now provided additional security for our own patrols.

We had a spate of speeding cars at night which used an effective trick of blinding our foot patrols with their headlights. We ruled 'side lights only' and I gathered that after some confrontations between rifle butt and headlamp glass this dangerous practice ceased.

I maintained this emphasis on the complete suppression of nightly activities until the moment we withdrew from Crater. That it was a successful policy is shown by the fact that in the entire five months of occupation the terrorist incidents at night were few and far between and in each case they were indirect attacks with mines or booby traps, with the notable exception of my own Land Rover. An important side effect of this was the fact that we could organize a reasonable amount of sleep for the Jocks not actually patrolling or on sentry, so that they were fit for the intense activity of the daylight hours.

The civil population, already thoroughly intimidated by terrorism, regarded the re-occupation of Crater as another convulsion in a life that was already familiar with violence and sudden change. Most of the communities were apprehensive, cautious and watchful, combining relief that the anarchy of the previous fortnight was over with a fear that we would take revenge for the events of 20th June. Many left for other districts of Aden, to return weeks later when their confidence was restored. But they refused to co-operate with us in any way. The Civil Police were hostile and resentful of our re-entry. As time went by, the presence of troops became accepted. It took ten days for the population to realize that the drastic reprisals they

194

undoubtedly expected were not, in fact, going to happen. The tough line adopted by the battalion earned respect. It also provided ample material for the smear campaign that sought to remove British forces from Crater by political means; since vigorous reaction to terrorist incidents demonstrated the will and ability of the Argylls to remain in Crater on our own terms.

Assassination, looting and arson had been commonplace incidents before the re-occupation. The escaped convicts from the civil prison were still at large. Trade was at a standstill, with the banks closed and many warehouses looted. There was a shortage of food and money. There had been no authority save that of the terrorist. With the opening of Main Pass to traffic on 6th July, taxis and buses gradually re-appeared. Crater was filthy, with refuse abandoned in every street, and normal sweeper and refuse collection systems did not operate effectively until mid-August. There was rumour of epidemics, and cholera was mentioned, but although there was a health problem it never got out of hand. Water, electricity and drains began to operate normally. Shops gradually re-opened, and day by day, as confidence was restored, more and more resumed an ever increasing business. This, though, was but a trickle of their former trade and stocks of most popular items were low.

After initial relief that the re-occupation of the town had been accomplished without casualties, there was an astonishing reluctance by the British authorities to appreciate that the military problem was still very real. Co-operation with the Police, though much exhorted, was a charade to which both sides paid lip service. At no stage was there any evidence that any of the South Arabian security forces were contributing to the suppression of terrorism. In the absence of any properly functioning Intelligence service I insisted on retaining a firm and absolute military grasp of the town.

I let it be known that if the 'Argyll Law' was observed this state of peace and returning prosperity would continue. All our emphasis was on keeping the peace and I became almost eloquent in explaining this through our Interpreter Team to any Arab or Asian who seemed remotely responsible. But if an act of terrorism were committed, I would give the order, 'Portcullis!'

'Portcullis' was a code-word I chose with care. It implied strength and security. This one word was an order that brought down the full weight of the Argyll and Sutherland Highlanders. With a crash like the descent of a portcullis at the gates of a castle, Crater was sealed. Road-blocks at Main Pass and Marine Drive stopped all traffic and within Crater nobody was allowed to move. Within a second or two of the explosion of a grenade or the firing of a shot, the Jocks had gone into action. Improvised road blocks sprang up. All cars were stopped. Meanwhile, the company in whose area the incident had taken place, raced into action. The Jocks spread out in the streets rounding up all terrorist suspects—usually males between the ages of 15 and 35—herding them to the compound we had erected for their reception. Within a few minutes all males who had been as near as three hundred yards of the incident were within the barbed-wire compound awaiting interrogation.

As each suspect was ushered or escorted into the compound those particularly suspected were isolated to one side. One by one, those most particularly likely to be terrorists were taken into Battalion Headquarters for questioning. Those led away doubtless thought that a terrible fate awaited them, a fear shared by all those in the compound. We did nothing to discourage this belief. Once inside the Chartered Bank, they were taken to specially constructed rooms on the ground floor, where they found Nigel Crowe or Brian Baty and his team of interrogators. Each suspect was photographed with a Polaroid camera and finger-printed and this, although it did not result directly in the conviction of a single terrorist through civil proceedings, helped dominate the area and cow our enemies and potential enemies. It greatly impressed the civil police who realized that it was the way that they should have been acting. Sometimes the questioning of a single suspect might last for more than an hour. Nigel, with his knowledge of the Arab personality, often produced useful information—although, as we knew, this would lead to no effective punishment. The questioning complete, the suspect would be driven in a closed truck to a distant part of Aden and told to make his own way home. If he really had a genuine reason for being in Crater he came back.

This was not bullying. It was the rigorous application of sound techniques learnt in a hard school. But, while it resulted in bewilderment and some apprehension amongst the citizens of Crater, it did mean that nobody was seriously hurt, let alone killed.

Having stolen the world's headlines it was inevitable that there should be considerable Press interest in us apart from the host of semi-resident journalists. A team from ITV's 'This Week', under Llew Gardner, were among several who came to do features on our exploits. Llew was told that he was free to go where he liked in Aden but that he should avoid going to Crater and doing anything on the Argylls as 'they'd had far too much publicity already'. It revealed an astonishing attitude and the result was a half-hour programme, twenty-two minutes of which was on the Argylls.

Worse still, behind our backs, the smear campaign was under way and they were not just good Moslems who joined it. There were some pretty dubious Christians in the act too. I only suspected this, but I was certain that for the sake of their reputation the terrorists would have to do something soon. We could not keep our guard up for ever. I knew that the peace must one day be rudely shattered. Our enemies had been shaken by our arrival but were now obviously sizing us up. Crater had not known a period of such calm since the beginning of the emergency, perhaps because our opponents thought we would tire of living amongst them and would leave of our own accord.

Finally on the afternoon of 15th July they struck. It was against Lieutenant Jamie Graham's platoon. He and his brother Euan, sons of General Freddie Graham, the Colonel of the Regiment, were platoon commanders in neighbouring companies. A grenade was thrown at a working party filling sandbags and Corporal Auchterlonie and three Jocks were wounded. 'Portcullis' was put into immediate effect but the terrorist responsible escaped.

The after-effects of this incident were not the sort of action any of my officers could believe although we had already become aware of the new and different campaign. Despite our success in holding Crater without deaths or serious injuries, complaints about our behaviour began to pour into the High Commission. That this was a 'smear' campaign, organized by the terrorists and the Egyptian Intelligence Service, seemed certain to me. But the complaints were not all from Arabs. There were British officers and officials who felt that we were alienating the civil population and increasing the dangers of a peaceful withdrawal at the end of the year. One rather moth-eaten person from the High

Commission had the impertinence to refer to my Jocks as 'Glasgow thugs'. I told him that 'poor white trash' such as he should be careful where he used such words in Crater and that my own mother came from Glasgow. I reminded him that he was talking about his fellow-countrymen and that they were worthy of his respect and admiration. He had the good sense to apologize and later became one of our strongest supporters among the 'White Arabs'. It is only fair to add that the most influential members of the British commercial community, led by Jock McNeil of British Petroleum, went out of their way to express satisfaction at the stability that our actions in Crater were bringing. One of them went so far as to send us an extremely generous three figure cheque on our return to Britain in appreciation of our assistance to his staff!

The campaign of denigration, however, went so far that, on 18th July, General Tower ordered me to use less forceful methods, or, as it was put at the time, to 'throttle back'. I was summoned to the RAF hospital at Khormaksar, where he was being treated for a bad leg, and he and I talked in the back of his car in the heat of the afternoon. I realized that to 'throttle back' meant that I must no longer order 'Portcullis'. I was to lose the initiative. There must be no violent rounding up of suspects; no raids on suspected terrorists' houses in the early hours of the morning; or searches of houses without the presence of Arab police—the surest way of blowing the gaff to terrorists I could think of. Altogether, I was to adopt a different attitude towards Crater.

I explained that this would greatly increase the danger. My Battalion would inevitably suffer casualties and so would the civil population. I had always considered my soldiers to be tough, intelligent, fighting leopards—not tethered goats. But Tower, although charming and sympathetic, was adamant that I 'throttle back', and gave me a sheaf of papers full of specific complaints against the Battalion, all subsequently proved unfounded. There was no point in arguing for he and I had problems of communication at the best of times. I looked at it from his point of view, remembered the 20th June, and said I would comply.

As I drove back to Crater I knew that I must tell the Battalion what had happened, what to expect and that, because our new orders went against all our training and methods of proved success, I must tell them that this was no change of mind on my part, in case they thought they were doing badly. It was four o'clock in the afternoon and it would be dark by six. The only method of ensuring that each of my thirty different positions and 700 Jocks received the news in time was by the normal method of internal communications within a unit, Part I Orders. I therefore quickly wrote and issued a special order on the widest distribution for the whole battalion to read. This said:

ACTION IN CRATER

1. The methods we have used to dominate and pacify Crater have brought a flood of complaints from Local Nationals and the Federal Government authorities. We have been accused of stealing, brutality, wilful damage and arrogance. This is the 'Smear' Campaign I warned you to expect. It was bound to come whatever we did and some of it was bound to stick. The fact remains that for two weeks we have preserved the peace in Crater with only three incidents at a cost of four of our own men wounded, one Arab killed and another wounded firing from the Aidrus Mosque.
2. I want all Ranks to know how much I personally admire the way the Battalion has operated since arriving in Aden. I believe that our techniques and methods have paid off. If the day arises when we can use them again I know that they are the answer to the problem. 1 A and SH is justly famous for its tough line and I know that you have all been terrific. Well done!
3. However, I have now been ordered to 'throttle back' in the interests of a political settlement. The civil population have squeaked and I am reluctantly forced to modify or abandon some of our techniques. I am well aware of the disappointment this will cause to all of you but I too am a soldier under orders and must be 100 per cent loyal to my own superiors just as you are to me.
4. Life will become a bit more dangerous now that we are prohibited from dominating the situation our own way. In the Argylls we thrive on danger so let us be even more alert—with fingers on the trigger for the good kill of terrorists which may soon present itself.

5. Company Commanders are to ensure that this order is brought to the attention of All Ranks in the Battalion by repeating it in Coy Details.   UP THE ARGYLLS!

That night, the 18th July, a terrorist was shot dead by one of our ambush parties near the spot where Corporal Auchterlonie's party had been wounded. The next morning Ian Mackay reported to me that the attitude of the civil police was definitely surly. I went round to visit the Superintendents of the Civil and Armed Police and got the impression that they were both a bit shiftier than usual. I suspected that, already, news that I had been ordered to 'throttle back' had reached the enemy through their friends among senior Arab police officers. This was apparent in their behaviour. Where they had been respectful towards us, Arabs were now arrogant. We soon showed them that they were optimistic and that, although we might no longer appear to be as all-powerful, we were just as tough and resourceful and just as ready as ever to kill a terrorist.

But life was to become more dangerous, as I had forecast. Two days later, the first Argyll since the 20th June ambush was killed by the terrorists.

The inability to be able to distinguish friend from foe is not a new problem in Internal Security situations. The loss of the Special Branch created additional problems in Aden where the local police force was heavily infiltrated, by both terrorist organizations, as of course was the South Arabian Army.

The para-military Armed Police, who had played the main role in the 20th June massacre, continued to operate from their barracks in Crater. On 13th July, we had to attend a ceremonial parade in their barracks at which General Tower took the salute. This was apparently to help to rebuild the status of the ex-mutineers—a somewhat hollow gesture when the parade ground was none too unobtrusively surrounded by a troop of armoured cars and upwards of a hundred Argylls, ready to shoot anyone not wearing a kilt—a dress I ordered for my officers that morning as a means of identification.

This strange farce of apparent co-operation was played through to the bitter end. The fact that the affiliations of the senior police officers were well known was studiously ignored by those above me. The first police chief in Crater was promoted out of danger for being too co-operative with us; the second was such an obstructionist that I got him sacked and the third survived because the Argylls guarded his house at the height of the interfactional killings.

It was the Jocks who found the activities of the police most difficult to understand. They knew that the police vehicle driving fast away from the scene of an incident was possibly carrying the mortar or men responsible but because of the rules we were not in a position to do anything.

This freedom of movement for the police and South Arabian Army, both by day and night, gave them ample opportunity to watch and report on our activities. It was no surprise therefore that our first fatal casualty since the re-entry should be in the post nearest to the Crater Police Station.

On the afternoon of the 21st July Lance Corporal Willie Orr was inspecting the observation post above the market place. Suddenly from the crowded tangle of buildings less than two hundred yards away came two bursts of fire from an automatic rifle. The shots, probably fired from the back of the room to hide the flash of the weapon, caught Orr where his body was exposed over the parapet and killed him instantly.

In the ensuing follow-up the weapon used was never found but one Arab tried to break out from the tight cordon, failed to halt when challenged, and was shot dead.

Willie Orr's death was a blow to everyone but I felt it particularly keenly as he had been a private soldier in my Company throughout the Cyprus emergency nearly ten years before. He personified the good Jock who serves for long years and is the backbone of the rifle section. He had been promoted Lance Corporal only a few days earlier because his active service experience gave him natural leadership in a dangerous situation. At the time of his death he was actually inspecting the section machine-gun to make sure it was in proper shape for the relieving sentry. If ever an Argyll died doing his duty it was Willie Orr. His example was to help me to do my duty now that weaker men were reverting to a policy of appeasement.

Terrorism is a dreadful weapon. Used against a colonial regime which is not fully supported by the Government at home or the administration abroad it can achieve the object of gaining peace at almost any price *and with no regard to the long term consequences.* This was the case in Palestine, Cyprus and Aden where the military hands were tied. It was not the case in Malaya and Kenya where sterner counsel prevailed. As I thought of Willie Orr my mind called up a quotation from Jean Larteguy's semi-documentary novel about the French Foreign Legion, *The Centurions*, which talks of 'Repressing those out-of-date notions which at once make Western man great and prevent him, nowadays, from protecting himself.'

But while I was engaged in this mournful soliloquy little did I realize that the crisis of my own personal position in Aden was reaching boiling point. Later that evening we were visited by Charles Dunbar and had a long and useful discussion about the conduct of operations. He was extremely helpful and contrasted so well with another officer who had written to me the previous day, 'It is quite obvious that the good humour that existed up to a few weeks ago between soldiers and local nationals has now disappeared.' The astonishing thing was that this was written exactly one month after the 20th June massacre and the officer concerned obviously believed that the Federal Government was going to take-over from the British on independence with a mutual exchange of 'good humour'.

Later that night, the 21st July, I was tipped off that I was required to be at the GOC's house the next day when I should be seeing the C-in-C. The advice given to me by the source showed some insight into my mood. 'Don't Colin, for heavens sake, tell him to go and . . .'

Unknown to me, during the previous three days while we were battling in Crater, at the cost of one Argyll for three terrorists killed, my Battalion Order of the 18th July had been bandied about at GHQ.

There was obviously no intention of disloyalty in this Order and, indeed, although there was neither precedent nor requirement for me to do so, I had despatched a copy personally to General Tower by hand when it was published so that he would know at once the Argylls were 'throttled back' as he had ordered. It was this copy that caused all the fuss, so I was told later. Apparently, Tower was reading a lot of papers in his home where he was confined with his bad leg. Some of them were to be passed on to Admiral Le Fanu and 'by mistake' the copy of my Order got attached to the bottom of the pile. When I heard

202

this account later I said, 'Anyone who believes that would believe anything!'

Next morning at nine, I had been summoned to the GOC's house on Command Hill. As I walked into his garden I saw General Tower on the balcony and said, 'Good morning, sir.' He replied, 'You know what this is all about. The ball is in your court.' I said nothing. Then, at the invitation of yet another of the many brigadiers in Aden, this particular one being responsible for disciplinary matters in Middle East Command, I was taken into the house. There I saw, seated at a table and wearing his naval cap, Admiral Sir Michael Le Fanu. 'Good morning, sir,' I said.

Tower was sitting on the window-sill nursing his bad leg. I remained standing in front of the table, like a deserter up on a charge before his battalion commander. Le Fanu did not look up but began reading from typed sheets of paper. I was never given or shown a copy but remember that it ran something like:

'Colonel Mitchell you have expressed views about policy in Aden which are contrary to those prevailing at present and in normal circumstances could be judged guilty of disloyalty and lead to your removal from Command of your Battalion. However, in view of your reputation with your soldiers and the quality of your battalion, to remove you would be extremely detrimental to their morale and this will not be done. But you must take this as a warning.'

Now Le Fanu is a man to look you straight in the eye, but on this occasion he did not once do so and he read the words without much conviction and in a voice that seemed to be concentrating on the letter rather than the spirit of what he said. I began to wonder what this was all about. Was it that the re-occupation of Crater had caught him unawares and he was under pressure from the Chiefs of Staff who would be getting stick from their political masters? I had thought it a bit odd at the time that immediately after the re-occupation of Crater the High Commissioner and the C-in-C had issued a joint statement confirming that their views were in complete agreement. Had there been some backfire with Whitehall? This seemed more like it than that he should be interested in my expression of the widely held view in the Services that the Aden policy was nonsense.

203

That was the end. I was not invited to say anything by way of explanation or to defend myself in any way. So I said, 'Sir!', turned and marched out, accompanied by the brigadier, who, once outside, told me, 'This is not a reprimand and will not be entered in your records. However, it will be kept in case the need to take further action arises.'

I said, 'Sir!' again—a useful word which, by the way in which it is spoken, can seem to mean many different things—from admiration to contempt.

As I drove back to Crater my mind was ice-cold and was quickly sorting out the implications of what had happened. *Point One*. My experience in Whitehall had convinced me that the lack of political aim in South Arabia was leading to a state of near anarchy. *Point Two*. I did not believe that the Federal Government would survive until Independence and, if it did, not for long afterwards. *Point Three*. My experience of 20th June had convinced me that the lives of British soldiers were less embarrassing to explain away than dead Arab mutineers and terrorists. *Point Four*. My personal view was that the military authorities were dragging their feet and far too concerned with the political aspects of Aden, when they should have been concentrating on the military problems.

Perhaps obvious disapproval of this was tantamount to disloyalty unless it helped put a bit of backbone into people and drew attention to the need for firm control. If this was the case, which I truly believed it was, it was wrong to consider me disloyal. I was directly responsible for the lives of 700 soldiers and for the safety of 80,000 people in Crater. I liked Le Fanu and he was patently as near to being a 'Big Man' as you are likely to get in the Services in peace-time. But he knew as much about 'the feel' of being an infantry battalion commander on internal security operations as I knew about battleships. However, he did understand men and he had the confidence of the High Commissioner and the Chiefs of Staff. If I could get him down to Crater to see things at first hand for himself he would probably have enough commonsense to realize that the Argylls were a first class show and our actions in Crater fully justified. I would never be reconciled to Tower. But then it was his fault, not mine, if there was a vacuum in the army command in Aden which I had filled somewhat inadvertently.

My inclination was to resign, now, in disgust, and get right out of that sort of army. I have always believed in 'back or sack'. But this would be deserting the Battalion at the worst possible time and I had experienced the effect on morale of a Commanding Officer leaving the Jocks under a cloud during operations. In addition, there was the vulnerability of the Regiment under the shadow of the wretched defence cuts and I was linked with that, whatever my personal problems. So, I concluded, the answer was that I should soldier on—'better men have fared worse'. Therefore it was in a grimly determined mood that I got back to Crater and, apart from warning the company commanders that my personal position was a bit delicate and that they should be cautious in discussion in front of anyone outside our own Regimental circle, I dismissed the incident from my mind and concentrated on the more important issue of defeating terrorism.

It was later frequently asked, both in and out of Parliament, whether I disobeyed orders in Aden. The question would have been better put if it had been whether *I obeyed* orders. I had instilled into my own officers and Jocks that the use of initiative is not disobedience. I therefore decided that, while I would have to use spectacular and effective methods of controlling Crater, in accordance with the GOC's order, I would quietly continue to run things my own way—as far as I could. I was convinced that I would be proved right in the end anyway.

Throughout that afternoon I drove around Crater to try and get the 'feel'. The company commanders all talked of mounting tension and the police were completely off the streets. Just before dark I held an Orders Group at the Chartered Bank. 'It looks as though we might be in for a rough night,' I said. About an hour later a contact report started coming in from 'D' Company. A patrol moving in an alley off Zafaran Road had surprised an Arab, who immediately grappled with one of the Jocks and tried to seize his rifle. In the ensuing struggle he was stabbed by the soldier's bayonet. Tom Kenyon, our Medical Officer, treated him in the Regimental Aid Post but the man died later. Had the soldier fired his weapon in the confined space, which he would have been perfectly justified in doing, he would have imperilled both friend and foe, yet his sensible action was obviously unpopular with those who did not remember that bayonets are issued to soldiers for close-quarter fighting and not as tin-openers. Shortly afterwards an order was issued prohibiting patrols from carrying bayonets. This had the effect of visibly reducing the deterrent value of the soldier in the streets. It was not, to my mind, a proper understanding of the principle of minimum force.

This incident was the beginning of a much more aggressive terrorist campaign against us and I took it that the terrorist realized we were not going to be removed by political pressure but were staying to punch it out for the full fifteen rounds. The following evening, Ian Robertson's company caught and shot a terrorist armed with an automatic pistol of Czechoslovakian manufacture, and small-arms fire was directed at our roof picquet on the Treasury building. The next morning one of our mobile patrols was grenaded and while the area was being cordoned off another grenade was thrown, which fortunately did not explode. It was a Russian RG4.

This pattern continued. I was not too worried about isolated grenade incidents or pistol shots but since Lance Corporal Orr's death four days earlier I had been thinking a lot about the problem of snipers. Then, to add to my apprehension, on the afternoon of the 26th July a sniper picked off another of our rooftop sentries, this time on the Gazeira Palace Hotel; fortunately the bullet only passed through his shoulder.

I had tried to hide any feelings of bitterness about the way I was being treated by the military management but, as always, in a close knit family team like ours, my friends knew me better than I realized. One night I found a letter on my desk from one of the company commanders which read:

> When the days we are living in are history it might be quite fun to look back over them and see what we all really did think and say.
> I would like to go on record, privately, in saying how much I, and everyone else, have admired the superb lead you've given us. We are all with you to a man—one only has to go round the Jocks to realize this.
> Having an inkling of the difficulties you have had to contend with from above and the problems you have yet to face reinforces everything I've said. You can, as I'm sure you know, count on every ounce of our support.

I fear I express all this badly but I know that commanding us is a lonely job, particularly here; I therefore thought it might help if I said what we all feel.

With the increase in grenade incidents, we were able to trace the pattern of terrorist action. The grenadiers had a simple escape route. They took refuge in their mosques. Crater, as the cultural and spiritual capital of Aden, had many of these scattered throughout the town, with tiny entrances between shops and houses sometimes barely distinguishable from a shop door. They were completely out of bounds to British troops—although there were precedents from internal security operations in India and Palestine for being allowed to enter in hot pursuit of terrorists. But our sole recourse was to ask for the South Arabian Army to search for arms and suspects and this they did without their British officers or advisors present. By the time they arrived and had given the premises a cursory lookover hours would pass and we knew that results would be negative.

Mosques were an ideal sanctuary. All the terrorists had to do was deposit a box of grenades in the mosque—which could easily be taken in by a police vehicle—and allow the grenadiers to choose their own target in their own time. We were powerless. This procedure was part of the farce. Nigel Crowe, who had served with and knew the Arab Army, told us we were wasting our time, but for the good of 'co-operation' we several times asked for Arab soldiers, though I can only remember two occasions in five months when they came. In each case their search yielded nothing. We knew the grenades were there. We knew the grenadiers hid there. We were like foxhounds waiting for the huntsman to get a shovel and dig out the earth. But the occupants were not game little foxes, they were murderers planning more murder—and they were getting off scot-free in the middle of a British colony while the Egyptian Intelligence Service paid them to be murderers and exalted them as freedom fighters.

On 27th July Admiral Le Fanu visited us in Crater. Five minutes before he was due to arrive at the Main Pass entrance a grenade was thrown from the door of the Aban Mosque, wounding an Arab boy walking in the street but otherwise exploding harmlessly against the side of an armoured patrol in Queen Arwa Road. I was waiting for the C-in-C so was unable to go and see for myself, but Nigel Crowe got the civil police along and asked them to investigate. They went through the motions of questioning one or two locals but refused to do more. I was glad that at last the Admiral could see our difficulties for himself and I said to him, 'This is just an exercise in self-control.' He replied, 'That's exactly what it is.' 'Or self-sacrifice!' I told him.

But at least the incident made our point about mosques, and the visit went well. He toured a number of our positions, was given a salute by the Pipes and Drums—Lochleven Castle—and was utterly friendly. It was the first I had seen of him since the previous encounter at the GOC's house and he told me that in reply to agitated enquiries from London he had signalled back confirmation that the conduct of troops in Aden had continued to be of most praiseworthy restraint. It was obvious that both he and the High Commissioner were backing our efforts and this was confirmed by Charles Dunbar, who had written to the High Commission refuting the 'smear' campaign. I had earlier written to HQ Middle East Command suggesting that a permanent civil affairs team should be attached to us. I had put Nigel Crowe on to visiting all the local dignitaries. One small boy offered him a 'smear' pamphlet with a big smile. It described us as 'Scottish Red Rats'. I sent an urgent message via Army Public Relations asking for the Scottish Press to come out to see for themselves.

But of course I had by now realized the vital importance of keeping our case before the public at home, because with nothing to hide I was only too pleased to let the Press and Television come to Crater and report what they saw. It was my best protection against 'smear' and good for morale. This was a fairly novel approach in the Army where the general attitude towards the Press is a defensive one. I found the correspondents a lively, intelligent and friendly bunch who were far more responsible than I had been led to believe. Although people accused me of seeking publicity for the Battalion I found it kept me on my own toes to deal with them frankly at all times. Gerald Seymour of Independent Television News got some particularly good action shots and he and many others added to our circle of friends.

Summarizing my views near the end of July, I wrote: 'We live in Crater cheek-by-jowl with the locals and I am informed that they respect us because we stand no nonsense. If we relax our vigilance for a second we get chopped. Provided we are 100 per cent on the job, twenty-four hours a day, I believe that law and order can be maintained. I am acutely aware of the responsibility we carry and the political implications and I appreciate the ramifications of what we are trying to do . . . perhaps we give an air of over-confidence which would be lacking if we had not trained so thoroughly. I think that this was the basis of a lot of misunderstanding as people found it difficult to accept that a new unit knew what it was about. I like to think that we are well past that stage of misunderstanding now.'

We began at last to pick up some Intelligence. A reliable source on 1st August told us that the grenadiers were all FLOSY terrorists while the NLF were doing the shooting. This made sense, particularly as the source also stated that both sides were impressed by our tactics and preferred to operate elsewhere unless a 'prestige' attack was necessary to support their extravagant claims against us. An Arab newspaperman told me: 'If the British Government had taken a tough line two years ago, as they are doing now at the start of the trouble in Hong Kong, Aden would not be in its present powerless, lawless state. But the incidents here in Crater are down by a ratio of 10 : 1 compared to two years ago. Assassinations of civilians are now NIL compared with a one per day average over the last year. This is due to your activities preventing the free movement of assassins.'

Towards the end of July the confidence of the terrorists increased as they found that we were hamstrung by their use of the mosques as a sanctuary and weapons dump. On the 23rd there were six separate attacks, but this time we were ready with a new technique and three terrorists were killed. Very quietly, in the early hours of the morning, we had put out pairs of our own snipers in positions covering the mosque entrances and with a clear view of the roads running past them. This meant a complete rearrangement of the OP positions but to conceal our moves we left life size dolls, 'dummy Jocks' complete with glengarries, in the existing positions and reinforced various key points with men who were normally employed on administrative duties.

The first grenadier attacked at 8.25 in the morning, throwing a grenade onto the balcony of a platoon headquarters. Fortunately it failed to go off, but at 9.49 and 9.58 our cordons were grenaded from either side of the market place.

This was the chance our snipers had been waiting for and as the grenadiers bolted back for cover, a distance of about a hundred yards from the mosque they were using, two single shots rang out from a direction they least expected. By three o'clock that afternoon a further three incidents had taken place and one of our soldiers had been wounded. But it cost the terrorist three dead grenadiers and although there were further attacks the next day I believe it taught the grenadiers that however sacrosanct their hideaway might be we had the answer if they emerged to continue acts of terrorism.

Any account of an operation always seems exciting on paper as incident seems to flow after incident in quick succession. But in reality there are often long hours, even days, of inactivity and sometimes boredom when vigilance wanes and concentration becomes difficult. The day-to-day routine, the administration, all has to go on and I as CO had to keep a sharp eye open to see how the company and platoon commanders bore up to the strain of these added responsibilities.

So July 1967 ended, with the terrorists coming out into the streets, confident at first that the old methods they had used so successfully in Aden would work against us, but gradually, with ten grenadiers killed, they began to realize that 'Argyll Law' was not to be flaunted. It was still early days, I knew in my bones that they would soon move in the hard men to try and dislodge us. I woke up on the 1st August, got dressed and sat on my bed testing the mechanism of my Browning automatic pistol—I had a hunch I was going to need it.

CHAPTER 13

'Nothing has ever been made safe until the soldier has made safe the field where the building shall be built, and the soldier is the scaffolding until it has been built, and the soldier gets no reward but honour.'

Crisis in Heaven by ERIC LINKLATER

THE TERRORISTS MUST have planned to move in their hard core 'specialists' for a 'D' Day on 1st August. At various conferences people had tried to make my flesh creep with rumours that 'the Argylls were for it'. 'Excellent', I always replied, 'that is exactly what we want—bring 'em out in the open.' But despite all the verbal shadow boxing I still got no real intelligence and I reconciled myself to listening to everything and believing nothing. I had one warning for the company commanders, 'All information is false.'

At five o'clock on the afternoon of Tuesday, 1st August, when I heard what was in fact the first 'bang' of the new offensive, I thought it was a grenade outside the Chartered Bank. I shouted 'Come on' to David Thomson who chased after me grabbing the self-loading rifle he invariably carried. But as we got to the door two more explosions told us that it was something different. It turned out to be two-inch mortar bombs. But it took us until first light the following morning to discover where they had been fired from. One of our patrols found the marks of a mortar baseplate in the soft earth and beside it three bombs which had obviously misfired. They were of Australian manufacture with an airdrop safety modification we recognized from Borneo days.

That was the beginning of the attacks on my HQ at the Chartered Bank. I gathered that in London some newspapermen who had been in Crater were laying bets that I would be dead by the end of the month. At the time the odds didn't bother me.

Our opponents now backed up their grenade and sniping attacks with mortar bombs, rocket launcher bombs, and anti-tank grenades. Ron Smith, my Quartermaster, had been the

Battalion assault pioneer platoon sergeant twenty years earlier in Palestine. He had been decorated for gallantry for bomb disposal work in Jerusalem and was to be mentioned-in-despatches again in Aden. I now got him to work as an unofficial 'expert' and he reconstructed the techniques being used in the rocket and projectile attacks so that our Jocks knew the sort of thing they were looking for. This speeded up our reaction to incidents and added to the general feeling of ascendency which I was determined to maintain.

The mortar teams achieved results of the standard of professional soldiers. Their weapon, when we eventually captured it, was a standard British two-inch mortar and the bombs from a batch presumably supplied to the South Arabian Army for their troops operating up-country.

At lunch time on the 4th August we had a number of visitors at the Chartered Bank who, either on duty, or out of simple curiosity, had come to spend the morning with us. The meal itself was over when the building was rocked by a series of explosions.

Three mortar bombs hit the roof a foot above our heads while three more straddled the building on the road outside. We raced through the door and up the narrow 'gangway' onto the roof. The piper on duty at the front of the roof was wounded but still at his post. On the ground immediately behind him was Corporal Jimmy Scott, the Pipe Corporal, who attempting to reach his wounded sentry after the first bomb, was struck by the entire contents of the second.

Some of the staff officers who had lunched with us had many years service but I know that what happened on that open roof top that Friday afternoon did more to bridge the gap between them and us than anything else could ever have done. It was always true to say that whatever my own differences with my superiors may have been neither I, nor any of my officers, had anything but praise for the work and the help which was given to us by all the branches of Aden Brigade Headquarters. Piper Oakley, the sentry, was later mentioned-in-despatches for his conduct throughout the whole time we were in Crater. Although wounded he continued to observe the flight of the bombs and described to me where he thought they were being fired from. We were thus able to get a fix on the approximate position of the mortar base plate and sent it over the wireless to our patrols. But we did not find it for some days. It was hidden in the Public Works Department vehicle yard.

My immediate reaction to the mortar attacks was to strengthen the defences of the Chartered Bank which up until then I had considered a lower priority than the platoon and section strongpoints. During the next few days so much sand was brought in through the doors that the Jocks renamed the building the 'Sand Bank'. Everyone back in our rear echelon at Waterloo Lines was put on to filling sandbags and moving them down to Crater. Dressed solely in shorts and assorted footgear it was impossible to tell what rank or position any of the hundred sweating filthy labourers held. One man did stand out, however, for he was carrying an old No. 4 rifle—a type no longer on issue to the Army. The Adjutant eventually stopped him and asked him who he was: 'Sir, thank God you've found me. I'm in the RAF. I was in your NAAFI when everyone was cleared out to fill sandbags. They won't let me get away—but I'd like to stay and help anyway now I've got this far!' He was a splendid airman.

In many ways this typified the atmosphere of Aden when the pressure was really on. The men of 60 Squadron Royal Corps of Transport, whose fellow drivers were massacred near Champion Lines on 20th June, carried jerrycans of iced drinks to our OPs up the steep hills around Crater before the re-entry. They were not ordered to, they were not asked to, they did it because it was their contribution to the 'war' and it was never forgotten by the Jocks on the hills. The sappers of 2 Troop 60 Squadron Royal Engineers, under Lieutenant Brendan O'Donovan, somehow managed not only to build defences, but made lifts work that had never worked before; fixed plumbing and solved each and every problem as it came along. In Aden the infantry for once stole the glory but this should never be allowed to obscure the part played by all the other arms and corps and, of course, the other two services.

On 3rd August two 94 grenades, fired from spigots, exploded near my Land Rover as I was preparing to go off to a conference at HQ Aden Brigade. They were fired electrically from a point a hundred yards away up Queen Arwa Road using wires and a battery. The next day five mortar bombs landed near Ian Robertson's company HQ and his escort vehicle was grenaded by a terrorist who quickly escaped. That same afternoon six more mortar bombs hit the Chartered Bank and later we wounded and captured a grenadier. Next day we wounded another in the market place after he had thrown a grenade at an armoured car passing the fish-market and later that afternoon, as I was holding a conference in my Tac HQ, a blindicide missile just missed the wall of my office. We put out a tight cordon and eventually found the firing equipment outside a building called 'The Modern Tower Store' which was an electrical supplies shop. The wires led into the shop so we arrested the two shifty characters who owned it and handed them over to the Joint Army/RAF Provost and Security Service. Later that night the Police rang me up to say that the shop had been smashed to smithereens by Arab looters.

Then our trouble with the mosques began again as the terrorists reverted to their old practice of operating from the sanctuary of holiness. Still, under the local ground rules, they were completely out of bounds to us.

As before, my answer to the problem was simple. I ordered the company commanders to site snipers to cover the entrances to all the most troublesome mosques. We did not have long to wait and soon caught a migratory grenadier at the Aidrus Mosque. But when we spotted a man armed with a rifle going into the Sayed Ahmed Mosque and requested the South Arabian Army to search it, the search, which lasted for four hours, yielded nothing. I was not surprised—just angry.

Throughout the long hot days of August the struggle continued. Slowly patterns of enemy activity were built up and as the days went by our successes mounted.

We learnt that though you could not always anticipate the first grenade you could almost guarantee that two or three others would be thrown within minutes of the first. This provided us with the chance we needed and time and again the grenadier responsible was cut down by an alert rifleman. The hours of short range target practice in Borneo, and later Plymouth, paid handsome dividends in the concrete jungle of Aden. There was time only for a quick almost instinctive shot but the knowledge that your target is trying to kill you can produce miraculous results from even the average soldier.

By mid-August our own information and intelligence sources

were beginning to work. From the day we entered Crater to the day we left I never received a single piece of worthwhile, practical intelligence from the official network. Even John Prendergast, famed from Cyprus to Hong Kong as one of the most outstanding Special Branch operators of all time, could do little but make prophetic guesses. 'Dicky' Bird, a forensic genius, left the scene soon after our arrival and with his departure even the most elementary investigations became impossible.

But Nigel Crowe and my Battalion Headquarters staff and intelligence section had been working overtime. Sergeant Allison, the Intelligence Sergeant, had served in Aden on detachment to special forces. Through friends in the SAA, suspects we questioned, contacts in Crater, and from our various visitors, particularly the Press, a small trickle of information started to arrive. In the best South Arabian tradition it was often confusing and contradictory but it gave us straws to grasp at. Money is the key to most Arab hearts and we learnt how to badger and wheedle to get our information. The amounts of money we distributed were trivial when compared with the risks informants took but eventually we had some success.

Our first major coup was a surprise source who told us the position of the main NLF press. This was in fact guarded by Arab soldiers as it was installed in the house of a Federal Government official immediately behind our headquarters. Although this meant we could take no positive action it gave us a lead to certain personalities and provided something of a lever. Before long we had a man translating all propaganda leaflets almost before they were printed and he secured a senior citizen who peddled a certain amount of propaganda on our behalf in his paper. At roughly the same time we received two welcome additions to our intelligence teams, Tony Milway, an Education Corps instructor from the Arabic Language School, and Captain Hamish Monro, one of our own officers who had spent two months in the Sultan of Muscat's Armed Forces brushing up the language. In the distinctive grey shirts which all my officers wore they spent their days questioning everyone and everything. Although we only had a maximum of six fluent Arabists in the battalion the word soon spread that all the grey shirts spoke Arabic, a most useful false rumour.

The pressure was now on the terrorist in more ways than one.

215

He got far less help than previously from the civilian population who were beginning to get confidence in 'their' soldiers. A story reached us from the Bazaar that the civilians had themselves disarmed a grenadier, so anxious were they to preserve the peace which the Argylls had imposed. By day he was harassed in the streets; by night he was penned indoors; and the dreaded grey shirts now had informers working for them.

The struggle continued. We had some narrow escapes and Brian Baty and the RSM escaped death by inches when they were dismantling a battery of primed spigot grenades as a grenadier lopped a 36 grenade from the Bank of India into their Land Rover, which caught fire. With burning vehicles, ammunition exploding, grenades banging and mortar bombs falling the battle was obviously joined.

Private Dickson of 'B' Company was another man in the wars. He was operating the manpack wireless set on 7th August when his patrol were grenaded. He shouted a warning and fell to the ground. The grenade landed between him and the patrol commander wounding both of them. Without saying anything about his own injuries Dickson radioed all the details of the incident and got the follow-up action going and medical assistance to collect the patrol commander. It was only when all was over and the patrol back in its base that he was seen to be in pain and admitted to being wounded by shrapnel in the arm and leg. It was serious enough to result in his being evacuated to hospital.

This typified the attitude of the Jocks—competent, cool, cheerfully devoted to their duty and generally uncomplaining. On 10th August an anti-tank mine was electrically detonated as one of our armoured patrols passed through the Forage Market. It blew a hole four feet wide and three feet deep so I assumed it was quite a giant. While the follow-up took place in the darkening, sinister shadows two men tried to break through our cordon and were shot. A sniper fired at Battalion Headquarters and next morning we found a bullet through the window of my bedroom, high in the top flat of the building. The Engineers put in steel shutters and I felt more than ever like a submariner.

Each day brought a mounting toll of incidents and we began to recover terrorist mines, explosives and grenades in the kutcha huts. These were all sophisticated terrorist devices, of Egyptian manufacture, with deadly electric detonators, time pencils and watches wired up for attaching to electrical detonator circuits. On the 31st August there were seven terrorist incidents and I turned in that night, well after midnight, wondering what the next day would bring. Our total Argyll casualties to date were five killed and eighteen wounded, while we had killed twenty terrorists and wounded five. Five civilians had been killed and twenty-seven wounded indiscriminately by terrorists. There had been one hundred and seventeen incidents since the re-entry into Crater.

Then, overnight, the terrorists lost heart. I believe that our finds of ammunition and explosives finally took the wind out of their sails. August, which had started like a lion, went out like a lamb.

The Argyll Peace had been achieved. For forty-two days from the beginning of September to the middle of October there was not a single terrorist incident in Crater. As day succeeded day the myth of Argyll Peace continued to grow. Shops that had been shut for months opened once more and life started to return to normal.

These forty-two days were a unique episode in the dismal Aden story. If the Argylls had done nothing else during their entire tour this period would have more than justified any methods that even the devil might have employed. In the quiet tranquillity of Britain it is difficult to imagine what normal life in Crater had been like for the past eighteen months. Murder, arson, robbery, violence of every conceivable kind had replaced the normal well ordered life of a British colony. Now peace had returned again.

Before coming to Aden I had trained my battalion against seven principles: Supervision: Self Control: Offensive Mindedness: Quick and Accurate Shooting: Well Tried Operational Drills: Intelligent Interest and Security Consciousness. I now had the consolation of knowing that training built round these principles worked. It was, to me, the justification of twenty-five years of soldiering.

Almost the only dramatic event in September was the burning down of the vast Crater market. This was a square brick and wood compound where the numerous stall keepers set up their structures daily for the seething mass of humanity that lived in the surrounding dwellings. Apart from fruit and vegetables for

217

the entire population of Aden, not just Crater, it was also the main centre for the qat sellers. This green leaf was chewed by the buyer and produced a drugged effect lasting for a number of hours.

The entire market area had always been a trouble spot and the inner precincts had been the scene of the only assassination to take place during our time in control. This took place under the virtual direct supervision of a police motor cycle patrolman.

Early in the morning of the 7th September it was another type of incident which drew attention to the market. Smoke was seen coming from one corner of the interior and within minutes the entire structure was alight. The blaze lasted all morning; in no way curtailed by the much famed but farcical Aden Fire Brigade.

It was not the fact of the burning that intrigued us, however, but the sequel. Before the day was out we were accused of having done the dirty deed ourselves. We had to submit to an official investigation that failed to break-down our cover story The reason was simple, we had not done it. The following day it was announced that the British would contribute the majority of the sum needed to rebuild the building. This was all the Aden Police had been waiting for. They admitted that a woman had informed them almost immediately of who had started the blaze and some of the culprits, all members of FLOSY, were already in custody.

As can be imagined the entire Battalion got much wry amusement out of this incident but in its way it was simply another indication of the multi-sided war we were waging.

I knew that eventually, for one reason or another, trouble would start once more.

The struggle for eventual power in South Arabia was slowly resolving itself. The cause of the Sultans, the original British hope, was now abandoned and the two surviving contestants were the main terrorist parties FLOSY and NLF. It is perhaps indicative of this entire period that FLOSY was at no stage made a proscribed party although the NLF was. Meanwhile the officer corps of the Federal Army split openly into supporting the rival factions. The Armed Police, whom I suspected to be pro-NLF, began to strengthen defences around the barracks and the next thing I knew was that they occupied two houses outside the perimeter of the barracks itself. This was too reminiscent of the scene before the 20th June so I went down and told the Superintendent that he would get all his men back inside by midnight or I would not be responsible. He reported to Said Abdul Hardi Shihar, who had come down to Crater on 9th October for an extraordinary conversation with me in which he explained the NLF case and seemed worried that if the Argylls moved out quickly the Armed Police, the NLF Commando, would not be able to cope with FLOSY. I said I had no intention of going anywhere in a hurry.

There was considerable interest shown by a number of senior officers as to how the average British soldier would accept any decision to recognize a leading terrorist organization. This to me showed a strange lack of understanding of the make-up of the Jocks.

When Brigadier Jefferies approached me as to how I would pass the news on to the Jocks I was slightly non-plussed. I did not see it as a problem; the rumours as usual had already told everyone in Aden, and I certainly had no intention of making a speech—or of issuing a Part I Order, for who knows where that would have found its way! I said they would simply do what they were told. They were professional soldiers, lean, mean and keen in the image of the age.

Meanwhile I was beginning to pose a number of questions to Aden Brigade anticipating the eventuality of having to intervene in interfactional fighting when and if it began in Crater. I first of all wanted to know the *Aim*. Was it to support the Police—in which case would we accept that this was tantamount to supporting NLF—or were we to be truly impartial and continue to stabilize the situation by the use of military methods without fear or favour. If neither of these, was it to be a 'ring-holding' operation, accepting the parallel of Palestine in 1948, when we permitted both sides to carry arms and establish recognized zones of occupation? I asked whether it was to be policy to leave the Armed Police in Crater as the NLF Commando—for that was virtually what the Commissioner of Police had admitted to me that they were. I wanted to know what would be the definition of an NLF/FLOSY supporter; whether or not the carriage of arms by civilians would be condoned; whether Crater could be sealed off with a 100 per cent search of vehicles and if my soldiers could be given the legal powers of constable.

I wanted to know: Would we declare martial law? How would we dispose of FLOSY prisoners/suspects assuming the Police (NLF) might murder them?

As I put all these questions to Brigadier Jefferies I sensed that it was only a matter of time before the British Government took the plunge and officially recognized the NLF. I was reminded of the cynical old story in my Regiment of the CO and the Adjutant in a difficult spot in war when the former said to the latter 'My boy! We must abandon this position—preferably with honour but if necessary without.'

Eventually, as the inter-party struggle continued, the British Government had to take the plunge and officially recognize the NLF. When it happened I reminded myself of what Nelson had said about the Neapolitans and how well the quotation fitted Harold Wilson and his Government: 'God knows they had little enough honour to lose, but what little they had, they lost!'

Once the Commissioner realized that I was going to have no nonsense over the Armed Police 'wearing two hats' and being NLF gunmen in the pay of Her Majesty's Government, he agreed to keep them inside the barracks except for routine duties. I said we would provide all the protection necessary for them—a strange quirk of fate for the perpetrators of the June massacre.

It was then that there arose, in October, what I came to call in lighter mood 'The Strange Case of the 36 Grenades—or who's on whose side anyway?'

Before we came to Aden there were in operation a host of restrictions on the employment and use of various weapons in the internal security role. It was one of these restrictions that had been invoked on the 20th June to stop the Saladin gun being fired. After our re-occupation of Crater I had clearly laid down that our isolated section posts should be as defensible as possible against any type of attack—by mobs of local nationals, terrorists, mutineering armed police or South Arabian Army. I delegated to my company commanders the decision of what they put where; but I personally deployed four WOMBAT anti-tank guns and the battalion 81 mm. mortars so that if street fighting ever took place, or we had to fight our way back to rescue our echelon in Waterloo Lines, the heavy weapons were under my direct control at various strategic points. Weapons and ammunition were always in ready-to-use locations and never hidden. I suppose that they must have been seen by many visitors during our five months occupation of Crater.

But one day in October when the Commander of Aden Brigade was visiting us I took him to one of Ian Robertson's posts in Salahaddin Road. The sentry had a small open cabinet beside him in the machine-gun emplacement and in it were three hand grenades, unprimed, but with the box of detonators arranged beside them. This was the normal practice in the 'A' Company OPs. The Brigade Commander commented that he might be wrong but he thought the Brigade orders were that grenades were centralized. I said no, my own orders were that 36 grenades were not to be carried by patrols lest they became classified as offensive weapons but it was at the company commander's discretion to site them defensively. It was the same with the Carl Gustav launcher. I thought no more of it but later the Brigade Major said that the orders now were for all grenades to be centralized not lower than Company Headquarters. I thought this was an unnecessary interference with a battalion commander's affairs but thought no more of it. However, some days later when the GOC was in my office I raised it as a question of principle. He was not prepared to discuss it. I held my peace, but quite soon afterwards in another Battalion's area an isolated OP was attacked by terrorists and subsequently abandoned amid casualties and street fighting. It had become indefensible.

Throughout the struggle in Aden one of the main objects of terrorist activities was to build up the prestige of the dissidents in the eyes of the local population. It was almost impossible to tell which of the various factions were actually responsible for an act, particularly when both groups' propaganda would claim the deed. A lot of the so-called terrorists were not full-time professionals but were simply paid to throw grenades or provide help for certain projects. As the power swung away from one faction this 'casual labour' moved across as well.

The climax in the inter-factional fighting was fast approaching and this made it all the more certain that 'vote' winning attacks would be made against the imperious aloof masters of Crater. This time the assault came in two very distinct phases.

First of all the NLF, always the more polished performers, brought in their 'official' mortar team once more. The first bombs fell on the 14th October and in the next twenty-four

221

hours there were four further attacks. None of the bombs caused any casualties and did no damage other than scar various buildings. As suddenly as they had started the attacks ceased.

Meanwhile Nigel Crowe took advantage of the rivalry to play one side against the other for information and in a raid we recovered anti-tank mines, grenades and various spigot-firing devices all set up and ready for use. We had a narrow squeak when a civilian car, parked and empty in the street, blew up just before one of our foot patrols went past. But for the good fortune of the wireless operator calling for a pause while he adjusted his aerial, the patrol would undoubtedly have suffered heavy casualties. Two nights later, at a time when our patrols were not in this particular location, a civilian Land Rover blew up near the police station and did quite a bit of damage to the surrounding Arab flats.

The Sultan of Lahej's Palace was looted of furniture on 13th October and we handed the culprits and their pantechnicon over to the civil police. It was difficult to know which were the more surprised—because the whole looting had gone on under the eyes of the police anyway and they obviously had an 'arrangement' with the thieves. The next day three cases of alleged theft were levelled against the Battalion. We carried out an investigation and fortunately traced the plaintiffs. They were three Yemenis who had been relieved of their money at Dar Saad, an up-country check point, and told to go into Aden and complain against the British Army. They had passed through the Main Pass road block at a time when one of my subalterns had relieved the platoon sergeant and was personally checking passes. I got onto the Commissioner of Police direct and said that the next time a 'smear' was organized they should do it more skilfully. He was very apologetic.

The last ten days of October saw some savage incidents. On the 20th one of our foot patrols was grenaded at nine o'clock in the morning, wounding two Arabs standing in the street and one of our own Jocks. The terrorist was chased and shot while another was wounded. Later the wounded man was seen in a taxi. 'A' Company had their road blocks out and the taxi was ordered to stop at the next corner. It drove on and the terrorist jumped out to be shot dead—as was the unfortunate taxi driver.

222

In 'D' Company's area, another grenadier was killed but his assistant was only wounded and escaped when they ambushed a foot patrol.

As I toured round between the incidents I turned to David Thomson and said, 'We're back in business.' At lunch time Paddy Palmer was grenaded in Aidrus Road and the grenadier ran back into the Askalani Mosque. I was incensed and told the civil police to clear the mosque or take the consequences. The grenadier finally emerged and was arrested and handed over to the Provost Services. He is probably a Hero of the Resistance by now. At 3 o'clock the last attack of the day was made when another foot patrol was grenaded in Zafaran Road. The grenadier was shot and later died.

It was a tense night in the Chartered Bank and around the section posts. We wondered what the next day would bring. A friend of ours in the British High Commission, Derek Rose, came to dinner. He was in good heart and we had a long and interesting discussion about South Arabia and the effect of Yemeni and Egyptian propaganda on the local people. Soon after dinner he left, promising to return in a few days. We were horrified to hear at lunchtime the following day that he had been assassinated in broad daylight by an NLF gunman on the waterfront at the other end of Aden. It was the beginning of a new outbreak of European assassinations.

At 10 o'clock the next morning I went round to the police station to get the 'feel' of things. As I came down the long flight of stone stairs in front of it I looked back towards the Market Place. Ian Mackay's Land Rover was turning the corner when a grenade exploded; one of his escort lay in the road and Ian himself, with drawn pistol, leapt out and gave chase. Corporal Mitchell and I saw the second of the two terrorists slipping off into the meat market and gave chase with David Thomson. We all split up and I found myself, quite alone, in the middle of a mob of Arabs. The grenadier had been wearing a particularly bright 'futah' and I thought I would recognize him if I saw him; suddenly, coming the other way through the dense crowd of Arabs, was David Thomson, also alone.

'What the hell are you doing here ! ! !'—we both said—but meanwhile 'D' Company had cordoned the area after Ian Mackay's party had shot the first terrorist, and the second was flushed out and shot seconds later. He had a Russian RG4 grenade still in the roll of his futah.

FLOSY had made a mess of their attacks and in these days, the 20th and 21st October, six terrorists were shot dead, one was arrested, two were wounded and two escaped. This was counter-terrorism at its best and was the culmination of months of hard work in Crater. The Jocks knew the area better than their opponents, many of whom had come in from other areas, and they were dealing with an enemy who now had little confidence in his ability to make a successful attack.

Of all the letters I received in Aden the one I got from Sir Humphrey Trevelyan on 21st October was the nicest, 'Well done —Good Luck'. It was small wonder he had our confidence, friendship and respect.

I reckoned that the harassing that they had been given over these two days would have taken the heart out of the grenadiers and I was proved right. For six days peace reigned. On the 27th a taxi broke out of the Marine Drive road-block when ordered to pull aside for searching. The covering sentry was unable to shoot because one of our own sandbagging parties was in the line of fire. Ten minutes later it was stopped and arrested by an armoured patrol but on the way back to Battalion Headquarters it again escaped. Shots were fired. The taxi crashed and the driver escaped unhurt into a mosque. The civil police suggested we kept the taxi until it was claimed by the owner!

We were now witnessing the start of the NLF take-over of power. The full story of this coup will probably never be known but there were some fairly dubious goings on. In a matter of days NLF swept the Sultans from office in the up-country states. Even the allegedly well entrenched Sharif of Beihan beat a hasty retreat into Saudi Arabia. The up-country take-over, although surprising for its speed and completeness, was not altogether unexpected. The main FLOSY strongholds had always been the urban areas and these still seemed secure. By this stage the British had already withdrawn from the outlying townships of Little Aden and Sheikh Othman and it was in the latter that the main battle was now to take place. Here in two days of heavy street fighting the South Arabian Army, now almost wholly aligned with the NLF, pulverized the main FLOSY terrorist band using all the weapons that our own troops had been banned from using

224

on the 20th June, including armoured cars with 76 mm. guns. By the end of the second day the main fighting was over although sporadic shooting continued.

What we witnessed during those days was perhaps the most cynical part of the collapse of British policies in Aden. The Federal Government, the puppets whom we had produced, had collapsed and we now had two factions openly fighting it out in an area for which we were still officially responsible. Like Pilate, however, we had washed our hands and were only too willing to pretend that nothing untoward was happening. Whatever may be said now, the NLF was not the Government of Aden and there were numerous hurdles to be crossed before the final withdrawal.

It was during those days that I was on my guard against any pressure to accept the activities of the NLF as official. The advances from the Arabs one expected—the suggestions and hints from elsewhere came as a nasty shock. In certain portions of Central Aden British troops had never really established control and were now only too willing to leave well alone. The *Sunday Times* depicted this situation extremely well under the banner headline:

ADEN TERROR: RIVALS IN DOOR-TO-DOOR MURDER CAMPAIGN. Rival gangs of Arab Nationalists took to the rooftops and back alleys again today to fight out to the bitter end the battle for control of Aden. Gunmen stalked the streets with lists of names of people to be killed.

Police officers were seized and shot. Bodies of victims were tossed from cars speeding through the narrow twisting streets. British troops kept out of it, with orders to do nothing unless Britons or British property came under fire.

'This is an extermination rampage' said one gunman. Each side —the NLF and FLOSY—is dedicated to the elimination of its rival by the time the British leave, later this month. In two days 50 bodies have been picked up in the streets; many more are thought to have been buried secretly. More than 200 people have been wounded, and an unknown number kidnapped.

Two senior Arab police officers have been kidnapped and killed —one at Steamer Point, the other at Little Aden near the BP oil refinery. One policeman was seized with his son: both were shot and their bodies dumped on their own doorstep.

The morale of the Aden police force is said to be on the point of

225

collapse, with officers in fear for their lives as the British with-drawal comes nearer. Many officers are said to be wanting to leave Aden as soon as possible to escape Nationalist retribution.

Civilians are already fleeing, some to neighbouring Yemen, others aboard crowded planes for Somalia, Beirut, Cairo and Bombay.

The fighting which broke out in the Sheikh Othman district on Friday, within hours of the British announcement of an earlier date for South Arabian independence, has spread to all districts *with the exception of Crater, where men of the Argyll and Sutherland High-landers dominate the streets* . . .

To me the whole affair was utterly disgraceful, particularly as these areas bordered the major shopping and shipping areas. Once British troops ceased to operate in these areas complete anarchy reigned, weapons were carried openly, interfactional murders were carried out in the streets and meetings could be held in complete safety as far as the terrorists were concerned.

At this time, as indeed on several previous occasions, there were suggestions that the extent of our patrolling in Crater should be curtailed. This showed a pathetic lack of understanding of our techniques—which I venture to suggest may become a copy-book for future IS urban operations. The essence of the domination method is to patrol constantly between the static posts. It is, indeed, only another interpretation of the basis of infantry soldiering—fire and movement. My line with my superiors was therefore straightforward even if not understood. I refused to allow what would be tantamount to 'licensed terrorism' in Crater. Any man who appeared in our streets carrying a weapon would be shot dead. To recognize the NLF was one thing: to expect the Argylls to support or condone the unlawful elimination of their opponents was quite another.

Obviously there were ways round our position, if for instance the arrests (in reality kidnapping) were carried out by uniformed police there was nothing I could do to stop them. Interrogations within Police stations could not be curtailed by my men but at least we could ensure that the eventual assassinations were not carried out on our streets. As Richard Cox, Defence Correspondent of the *Daily Telegraph* said, 'As the current fighting shows, FLOSY and the NLF are lined up against one another. Only in Crater, where the Argylls deserve great praise for keeping the situation in control, are they frightened of coming out into the open.'

Our stand had two very interesting sequels. The announcement of the recognition of the NLF led to the most amazing display of party strength yet shown in Aden. Every car, bus and building was immediately decorated with the red, white and black flag of the NLF, and this included all official buildings such as police and fire stations in the middle of British controlled areas. Only in Crater were they absent. The sole flags seen were the Union Jack, the green and yellow banner of the Argylls, and the official police ensign. At the two entrances to the town each car or bus stopped and the occupants carefully removed their flags and stowed them away until they left again. I continually made the point 'We are impartial. We shall shoot anyone who practices terrorism—that is what being impartial means, in Crater.'

During the heavy fighting in Sheikh Othman the casualties soon flooded all available hospital facilities, and then came that other burden of modern times, the refugees. As usual they ran for the nearest available sanctuary. *The Times* of 8th November stated, 'By and large, however, the Argylls' reputation for fast and true marksmanship seems to have been bruited about. During the fighting refugees were pouring in to seek asylum in an area which was once the focal point of internecine war and remains one of the last areas policed by the British Army.'

Our next contact was not until late in the first week in November. Early in the morning two shots and an explosion were heard in the north end of the town. A Mercedes taxi was seen making off at high speed but the patrol that investigated found nothing of interest. About fifteen minutes later a foot patrol saw a Mercedes taxi trying to make its way against the traffic up a one-way street. As they moved to investigate one of the patrol noticed that one of the passengers had a grenade in his hand. He shouted a warning to his companions as the four men leapt out. The driver escaped but two of the occupants were shot dead there and then, one of them carrying a ·38 pistol. The grenadier, wounded in the first exchange, was pursued down an alley-way and was eventually killed in the next street. Police sources later claimed that all these dead men came from a FLOSY commando which normally operated in Sheikh Othman.

The following day, the 6th November, was a somewhat

227

poignant one for all as the battalion said goodbye to Nigel Crowe. He had joined us in Palestine in 1948 and had soldiered on through the years to receive no reward except nine bars to the two General Service Medals—which I always pulled his leg about saying that it must be more than any other officer in the Army. He was a proper 'bushwhacker'. At the time there was much speculation about his early departure and until now the reason he left us two weeks early has never been told. Abdul Hardi, who had become the first Arab Commissioner of Police, was scared stiff of the contacts that Nigel had with the South Arabian Army. More than once when I asked his opinion he had replied, 'Why ask me? Major Crowe has far better sources than I.' As the pressure built up and Abdul Hardi was making his own bid for power, he became more and more worried. Eventually, on one of his clandestine visits to my Headquarters, he said, 'It would be dreadful if anything happened to Major Crowe.' He seemed blissfully unaware that after his previous visit David Thomson had jokingly said to me, 'The weapon that would make that guy yet another Arab martyr needed only a nod from you for the Jocks to execute the deed!' I made it quite clear that I was not in the habit of having my officers threatened and that I would not feel responsible for the consequences of the retribution if a hair on Nigel's head was touched. By this time, however, the struggle within the South Arabian Army had reached a stage at which even I thought it somewhat risky to allow him to go on visiting SAA camps and houses and I therefore curtailed his movements. When the chance came to get him home early I put him on a plane which if nothing else gave him an extra fourteen days to arrange his official retirement from the British Army. He was an East of Suez man.

Having said goodbye to Nigel, I went back up to the Mess in the Chartered Bank and started to play bridge, my conventional relaxation on a slack night. We did not get very far. Soon after 11 p.m. Ian Mackay came in to report some shots that had been fired in a street about 300 yards away. Someone had fired at the guard on the house of Thaabit Mohsin, the Superintendent of the Crater Police, and while the police returned the fire 'D' Company cordoned off the area. During the search that followed twenty-seven men were found sleeping in a building still under construction. A British made 36 grenade was found under one of the beds and as the owner ran away he was shot and killed. The police were now getting stuck into the other occupants and eventually claimed that four of them were from the same FLOSY commando that we had tangled with earlier. These men they removed and were not seen again.

By the time Ian had finished his report, my other company commanders had also been in touch with me and they all felt that there was something in the air. In view of this I decided that I would go and take a look round myself, first visiting Ian Robertson's company behind the Armed Police barracks.

As I drove up Queen Arwa Road it was within half an hour of midnight. I felt that thrill of expectation which makes soldiering such a worthwhile profession. We were on the verge of some sort of excitement but I could not yet define it. Oddly enough the very last thing I expected was what then happened. As we sped past the long wall of the Armed Police barracks, illuminated by the arc lights of Queen Arwa Road, a grenade came flying over the wall out of the Police Barracks and landed behind my Land Rover. It exploded beside my escort vehicle and Brian Baty and two of the Jocks got fragments of shrapnel into them. We were only a hundred yards past the spot where Bryan Malcolm and John Moncur had been massacred on the 20th June. Now, on the 6th November, after exercising more self-control and restraint towards the Armed Police than their treachery should ever have merited, they had tried to repeat their murder. They will never know how lucky they were that night. There were very detailed orders from Aden Brigade about action to be taken in the event of hostile acts by the South Arabian Army or Police.

But it is the immediate action of the man on the spot which matters and something had to be done very positively. I decided, as the Security Commander of Crater, to sort this out myself rather than invoke the various code words and procedures which would have caused delay and indecision all along the line. I pulled in behind the cover of some houses and we got the two Jocks out of the road where they were lying. Brian was only wounded in the hand and bound himself up with a handkerchief to stop the flow of blood. Over my wireless I ordered Ian Robertson to surround the Barracks and told the Duty Officer in Battalion HQ to warn the Armed Police that I was coming in and wanted to see the Superintendent immediately.

I then started to walk, in full view of the Armed Police sentry towers, down the road back towards their gate. An armoured car and a section of Jocks from 'A' Company caught up with me and we must have looked a pretty formidable party by the time I reached the front gate. I went in with David Thomson, Brian Baty and my escort. The Police, in various stages of undress, were queuing up at their armoury to get their weapons out.

I quickly identified their Superintendent, Ali Gabir, and told him that he should get the Commissioner of Police on the telephone and tell him what had happened. Ali Gabir was horrified as his interpreter told him what had happened. I did not know whether this was because I had been ambushed or because the grenade had missed me but I suspected the former as he was well aware of the weight of fire power we had lined up for just such an occasion.

I walked him through the Barracks, across the parade ground and up to the wall from where the grenade had been thrown. He expressed horror, surprise and dismay in a gabble of Arabic which everyone was trying to interpret his own way. I felt fairly certain that he had not planned the grenade attack and that the majority of the armed police had no idea what was going on.

I spoke on the telephone to the Commissioner of Police who apologized most profusely, and said that he would carry out a full investigation immediately.

So, in the dark shadows of the Armed Police barracks in the early hours of Tuesday, 7th November I quietly lectured the Armed Police on the narrow escape they had had. I pointed out that we would have been justified in shooting back and that it was only the discipline and restraint of my soldiers which had prevented a repetition of the 20th June—except that this time *they* would have been lying dead.

I believe they got the point because by the time I left Ali Gabir they were all full of profuse apologies and obviously so relieved you would have thought it was a convocation of bishops. I then left the Armed Police Barracks for the last time.

Next day the Commissioner of Police telephoned to explain that the grenade had been thrown by a FLOSY recruit in the training company who had been arrested and removed from Crater. I shall never know the truth. Perhaps they had rigged the whole incident to try and get FLOSY blamed for my assassination so that they, the NLF, would be allowed to sort out FLOSY in Crater. What was certain was that by taking prompt action we had retained the initiative and stopped them making any capital out of it. I did not subscribe to the view, expressed later, that I should have kept out of the Police Barracks in case it sparked off the Third World War. The point was, as always, that firm action by the man on the spot is invariably successful, provided he believes in himself.

The following morning we had perhaps the most dramatic incident of all. Just after half past eight one of our foot patrols encountered a car with a suspiciously large number of young men in it. As they approached the car stopped and the occupants leapt out. Three of them surrendered immediately but the other four, armed with pistols, decided to shoot it out with the Jocks. In a scene which resembled a fancy-dress version of any western cowboy film, the Jocks gunned them down. All four were killed before they could inflict any damage at all. In the investigation that followed it was discovered that one of the men who surrendered was a Shell employee who had just been kidnapped. His abductors were in the process of taking him out of Crater to execute him when they ran into the patrol.

Ever since the first mortar attack the one man we had wanted to capture more than any other was the operator of this weapon. The marks on the base of each bomb told us there was only one weapon in use and the skill of the user pointed to it being the same man each time. Then on the 11th November the mortar fired again. Two bombs missed all military targets and wounded two Arab children. A mobile patrol was almost on the spot and pounced immediately. The mortarman fled, but the weapon, damaged by the hard ground, and one bomb, were recovered. This time, however, the terrorist was separated from his getaway vehicle and was picked up by another patrol within minutes. He was handed over to the provost services, to be released a few days later when the British left. To add insult to injury, the police later came to us and handed over two further bombs which they 'found' in the area of the base plate position.

The interesting thing is that *Life* magazine had a team in Crater which we agreed to escort round and they photographed both the fall of the bombs and the capture of the terrorist. Someone said to them, 'A typical bit of Argyll propaganda'. Then they were shown the photographs of the incident and the wounded Arab children.

The days that followed were tedious in the extreme. Although there were no incidents we could not relax as the best time to hit your opponent is always when he feels that he is home and dry. To the Jock it was the usual story of rumour and counter-rumour.

Our own preparations for the evacuation were simple—except for Ron Smith, the Quartermaster, and his staff. We had returned all non-essentials some weeks before and were virtually living off our backs. The South Arabian Army toured our positions, to my mind a needlessly dangerous performance, and were shown whatever they asked to see. Their real concern was that we should not remove the valuable 'loot'—refrigerators, air-conditioners, carpets and other attractive items, as they obviously planned to repeat in Crater the wholesale sackings they had carried out in Little Aden and Sheikh Othman. The charade ended with a buffet lunch and speeches, translated from one language to another, in the Mess. It was a form of hypocrisy which was not to our taste.

From the point of view of a battalion commander, the withdrawal posed few problems. If there was no opposition we left by the Royal Air Force Air Support Command aircraft from Khormaksar. In case the situation deteriorated and proper fighting started, we had accumulated considerable reserves of ammunition, removed our already small administrative tail, and were prepared to stage a fighting withdrawal to whichever pick-up point the Naval armada wished to use. The one thing I was determined to do was to ensure that no man had to hand in his weapon before he left.

The final withdrawal produced nearly as much acrimony for us personally as the retaking of Crater had done five months earlier. All the planning had been done on a 'W' (Withdrawal) Day basis. It was, understandably, meant to be terribly secret and no announcement was to be made as to when 'W' Day was until the last moment. But it was not hard to guess because, like many sound military plans before it, the administrative preparations had already given much of the tactical plan away to anyone reading between the lines.

Briefly, the plan was for the three units still responsible for

232

Arab areas, 42 Commando Royal Marines in Tawahi, 45 Commando in Ma'alla and ourselves in Crater, to withdraw simultaneously in the middle of the night. The Argylls were to leave by air almost immediately, followed by the 1st Battalion The Parachute Regiment who were holding the northern end of the Khormaksar plain. The two Royal Marine Commandos were to take over from the Parachute Regiment in the north and occupy the key features in the south, including the lip round Crater, thus securing the airfield until the last moment. 45 Commando would then depart by air leaving the last unit of all, 42 Commando, to hold the ring and withdraw by helicopter to the Commando Carrier HMS Albion lying offshore. As far as we were concerned the plan was to withdraw on the night of the 25th/26th November.

It was a quiet day because the Press, of whom there were a very large number come to report these closing moments, were attending a Fleet Review.

The Royal Navy quite rightly regarded the withdrawal from Aden as a tremendous opportunity to publicize their role as a Task Force and had collected together a truly imposing concentration of ships. I had accepted to go as a spectator to the Fleet Review but had to cry off because that night was to be our final one in Crater. During the afternoon I told the company commanders the move was on and outlined the details of my plan. They showed no surprise—there were too many other straws in the wind to keep it a secret. I gave it the codeword 'Highland Clearance'.

Public Relations had made a plan to divide the Press Corps into three groups, each to go to one of the units withdrawing. As the timings of all three units would be simultaneous there would be no point in moving from one area to another. It was a neat and tidy military plan. They had been summoned to a conference at 10 p.m. to be given the outline of the evacuation plan and to allocate them to their respective groups. But many experienced journalists, who knew exactly what they wanted to report, did not even bother to go to the Press Conference. They came direct to Crater.

As the tactical withdrawal started I placed myself at the Battalion Checkpoint, just outside the Chartered Bank, through which everyone had to pass and from where I hoped to be able to control the whole operation and watch the progress for myself. I was determined that there should be no hitch in these closing moments.

I hoped I was geared up to cope with the Press and that the small ceremony of lowering the flag at one o'clock in the morning would provide them with all they needed. But I was wrong. There were seven different Television teams in Crater that night including one from the United States and another from France. I toured the streets of the town with David Thomson and my escort and was convinced, from this and the reports I received from the company commanders, that intervention by terrorists at the eleventh hour was unlikely. In any case my plan was tactically sound and it would have been almost impossible for any terrorist to break through the Highland Line, as I termed our defensive cordon.

In the midst of all this General Tower arrived. He took in the scene at once and realized that almost all the Press Corps were there. He asked one journalist why he had come to Crater and not gone to Ma'alla. It was a pity that it had all worked out that way because we were leaving a town that we had fought to re-occupy and control in our own way and where five Argylls had died and others been wounded.

Suddenly I felt terribly weary of it all. General Tower had never understood our problem. I told David to arrange for him to be sidetracked away from us. It was impossible to conduct operations with a disgruntled GOC and hordes of Press and visitors, however charitable and confident one was.

The withdrawal went according to plan, culminating with the final act of opera when we physically handed over to the South Arabian Army a box full of keys to every post, house and flat we occupied in Crater.

I left at the last moment, passing through the platoon of 'B' Company who as the first to enter Crater were to be last to leave. As I did so I sent over the Brigade Radio Net the codeword to indicate that we were successfully clear of Crater. It was politely acknowledged with a congratulatory message to which I replied, 'Up the Argylls'.

We were now officially non-operational and all we had to do was report to the airfield at the appropriate time. I was due to leave on the first aircraft at 6.45 a.m.—in about five hours time.

Our fly-out plan had been agreed by everyone but, inevitably, was subjected to the intervention of the GOC who for some motive I did not understand at the time switched me to a later aircraft. The reason was apparent when I reached Bahrein and was immediately shown a copy of a signal to the Director of Public Relations in the Ministry of Defence alerting him personally of the time of my return so that I could be met on arrival. Little did I know it but this set the pattern for the next few months. My days as a soldier were already numbered.

From Bahrein we were flown back to Wiltshire in a Britannia aircraft of the Royal Air Force. As we sped through the air my mind went back over the six months that had elapsed since I first set out from Gatwick with the advance party. What were the conclusions from our tour in Aden? What was left for experience after you had sifted through the ash bucket of heroism and ineptitude, goodwill and petty jealousy, life and death which had made up the Aden we had known? As a professional soldier I knew that my 'Seven Principles' of internal security were valid, but there was far more to it than that.

One could run over all the other military lessons, too, and reach a number of valid and indisputable conclusions for the textbooks and pamphlet writers. But these were less troublesome than the fundamental issues. It was said that the Argylls had an easier job than other units because we had fewer incidents. But I regarded 'incidents' as a lack of military efficiency and was ruthless with myself and others in suppressing them, because they were indicative of that apt remark of Wavell's to Churchill: 'Big butcher's bills are not necessarily a sign of good generalship'.

It was also said that I arrived in Aden with preconceived ideas about how to do things and a juggernaut attitude to opposition or advice from those above me. But the facts were that I arrived still with an open mind and the tragedy was that my superiors never gained my confidence, and indeed after the 20th June I feared more for the lives of my men than the approbation or approval of authority. It was acceptable that they had no confidence in me either and indeed they said that I was rigid in my outlook and showed no understanding of their wider problems. But there was method in my madness and I not only understood the wider problem but I purposely and with intent created a

235

political factor, the stability of Crater, which the High Commissioner told me became a major factor in effecting an orderly withdrawal in hopeless political conditions. It was said that I was a publicity seeker and, I gather, one critic described me as a 'bloody little charlatan'. But I took flamboyant personal risks in order to demonstrate to my own officers and NCOs that we led from the front and my critics never appreciated that I was taking more risks on the quiet than those they viewed or read about. In terms of publicity I was completely honest with the Press and was never once let down or exploited by them. The military authority could browbeat me because of their rank but they could not browbeat the Press, although they frequently tried to do so. This put me in an impossible situation as 'the jam in the sandwich'. It was said that other units suffered because the Argylls got most of the publicity. But this assumes a lack of credibility about the press correspondents which no responsible observer could sustain; indeed it was policy to divert publicity to other units and prevent it from going to us—but this typically socialist doctrine of equal benefits for all could never work with the representatives of a free press. It was said that I was ruthless —but never by anyone who had the evidence of his own eyes to record the events and casualties of 20th June or who considered the lives of British soldiers an unsuitable currency to pay for the rewards of political expediency. It was said that I practised counter-terrorism rather than duties in aid of the civil power. I did—what 'civil power' was there in the accepted sense?

But these conclusions were of secondary importance to the main lesson learnt from the tour of the Argylls in Aden—the strength of the Regiment. I had written earlier that we were a band of brothers and not a flock of sheep, but we were more than that. We had mostly known each other for many years in good times and in bad. I knew the strength and the weaknesses of every officer and NCO in the team and had seen how they reacted on active service. Furthermore, being a family Regiment, we all shared similar interests and memories. This applied equally to both officers and Jocks many of whom had joined because their fathers, brothers and uncles had before them, or they identified themselves with a great fighting tradition. I knew that the Argylls could not have done the job they did in Aden if we had been a numbered unit without tradition and without a soul as some people in Whitehall would try and have us. The wonderful thing about soldiering with the Battalion was that when things got most difficult we closed our ranks and fought on. We sustained each other.

CHAPTER 14

'We can make a catalogue of the moral qualities of the greatest captains but we cannot exhaust them. First there will be courage, not merely the physical kind which is happily not uncommon, but the rarer thing, the moral courage which Washington showed in the dark days at Valley Forge, and which we call fortitude—the power of enduring when hope is gone, the power of taking upon one's self a crushing responsibility and daring all, when weaker souls would play for safety. There must be the capacity for self-sacrifice, the willingness to let worldly interests and even reputation and honour perish, if only the task be accomplished. The man who is concerned with his own prestige will never move mountains. There must be patience, supreme patience under misunderstanding and setbacks, and the muddles and interferences of others, and the soldier of a democracy especially needs this. There must be resilience under defeat, a tough vitality and a manly optimism, which looks at the facts in all their bleakness and yet dares to be confident. There must be the sense of the eternal continuity of a great cause, so that failure and even death will not seem the end, and a man sees himself as only a part in a predestined purpose.'

Homilies and Recreations by JOHN BUCHAN

WHEN I BROUGHT the Argylls back home from Aden at the end of November 1967 I found myself in a situation which was as disturbing as it was unique. I had become something of a national hero with the droll nickname of 'Mad Mitch'. The Argylls were a household word, quite rightly in my view as it was high time the Argyll Jocks had a bit of publicity for their years of loyal service to the Crown. They were the salt of the earth.

Personally, I assumed that despite my differences with higher authority in Aden I would get a square deal from the Army. I was keen, ambitious and filled with a love for soldiering. I longed to get command of an infantry brigade and put all my ideas on training and organization into practise. I was now forty-two and before long would be too old for proper field soldiering, which I had always regarded as a young man's game despite the

238

tendency in peace time to have senior officers about five years too old for their jobs.

I regarded the Aden affair as water under the bridge. I would plug on. Then the *Royal United Services Institute Journal* published an article about South Arabia in which the author made some unfavourable comments about the Argylls in Crater. I was furious about this as we had most diligently kept quiet about our side of the story and I had a pretty shrewd idea who might be behind this unchivalrous attack on the honour of the Regiment. After talking to my old boss, General Sir Gordon MacMillan of MacMillan, I withdrew the scorching reply I had sent to the Editor in favour of a milder letter from Sir Gordon in the role of an 'elder statesman'. Quite independently, Tom Pocock, the Defence Correspondent of the London *Evening Standard*, waded in with a most complimentary first-hand account of our activities. He had been in Crater before and during our tour and knew all the personalities involved. He compared it with the Casbah in Algiers. His sense of fair play was outraged by the treatment we were receiving and he made the point very clearly, following this up with a very good newspaper article explaining the Argyll fight for survival and my personal predicament.

The next bit of irritation was a public outcry because I was not awarded the Distinguished Service Order like other CO's. This arose after a very generous list of Honours and Awards for Aden had been published, which gave me a lot of quiet chuckles and combined with the New Year's and Birthday Honours for 1968 to confirm my long felt cynicism of the whole system. I told some of my more eloquent supporters: 'Belt up. The Argylls won six VC's before breakfast at Lucknow.' It was embarrassing and irritating but of no consequence.

The most important issue to me in January of 1968 was the survival of the Regiment—and there was every inducement to fight for that with a prospect of success. The Government were spending vast sums of money trying to attract recruits into the Army and the Argylls were the best recruited Regiment in Scotland: even a Socialist Government could understand the economics of that. In May, I took the entire battalion to Scotland and there we paraded and recruited in our Regimental area, obtaining even more recruits. By any practical standards our future should have been secure.

239

In terms of my personal career the test of the Army's intentions would come when my name came up for consideration for promotion to full colonel or brigadier. I could reasonably expect advancement, because the very few who held brevet rank were 'groomed for stardom'. But I was informed privately that I was to be sent back to do another first grade staff appointment, such as I had had before taking command of the Regiment. The reason given was that in Aden, despite earlier recommendations that I was suitable for command of a Brigade which had been accompanied by the rare gradings of 'Outstanding', I had been removed from the list on the recommendation of Jefferies and Tower.

I knew too well the workings of the military mind and the Whitehall machine. There was now no worthwhile future for me in the Army. For years I had fought against the caste of senior officers who were essentially bourgeois in outlook. I knew their weaknesses too well; they feared men who shared their better values but who despised their lesser vanities. Their ideas and tastes too often reflected a world of thirty years before; they pretended to come from a governing 'class' and yet they had no concept of leadership—they never took a risk unless someone else could take the responsibility for failure, they were utterly devoted to conformity and 'playing it safe'. I was frustrated by this narrow world of Army politics and although there were so many very good men, at all levels, I wanted no more of it.

Just as I had hung on in Crater I was anxious not to make a move until the decision was reached about the future of the Argylls. But as national speculation seemed to centre round my personal future I began to feel that my presence in Command of the Regiment would make the Whitehall warriors even more unsympathetic to our cause. I therefore thought this the best opportunity to make it clear that I was resigning for purely personal reasons—as if the reason really mattered anyway. Anyone who knew anything about the military 'Establishment' knew exactly what had happened in Aden. I believed in myself and my methods. I had no time for senior officers who had failed to spend the years since 1945 in the pursuit of active service. I had proved my superiors wrong on every count. I aroused bitter jealousy among less publicized units and above all I was successful. In the Britain of the nineteen sixties success was not considered a military quality unless it was accompanied by that sycophancy and mock modesty which placated the old men of the tribe.

Once my future was disclosed I sat down and wrote by return a letter to the Ministry of Defence, submitting my resignation. Curiously, the date they had chosen for their letter to arrive with me was the 20th June 1968—the first anniversary of the treacherous massacre of twenty-two British soldiers in Aden and what I personally considered the failure of senior British officers to save their soldiers. It was wholly appropriate that I should retire on that day.

The future of the Argylls remained unannounced. Our recruiting record had risen to the top of the Highland Brigade and we were rated one of the most efficient battalions in the British Army—if not the best. Certainly we were the most experienced operationally. But all our efforts had no effect. The faceless committee men in the Ministry of Defence, the 'gnomes' of Whitehall, were determined that the Argylls must go and the decision was announced in July with a suitable display of crocodile tears. There was an inter-party debate on the matter and with dreary finality the Government Ministers defended their action by sheltering behind the decision of the Army Board— another Committee with the distinction of having no Scottish representation on it whatsoever.

Faced with the need to create a diversion from the real issue, the Labour Party managed to dig up a rather unlikely socialist Member of Parliament, of ancient Scottish lineage, who launched an attack against the Argylls in Aden in which he implied ill-discipline by the Battalion and disobedience of orders by me personally. This was an impossible situation for me as I was still a serving officer and the attack was launched under Parliamentary privilege. Although Healey, the Defence Minister, ordered an inquiry I was not asked or allowed to give evidence. I therefore decided to write this book.

Of all the futile acts of self-destruction proposed by British Governments during those tragic years before the 1970s, nothing personified their death-wish more than the decision to abolish the Argyll and Sutherland Highlanders. The Argylls were a symbol of all that was best in the British race. They had campaigned for decades with quiet purpose and their quality had brought such natural applause that it is small wonder they became the heroes of the British public during the fatal months preceding the evacuation from Aden in the winter of 1967. Here, indeed, was a great Regiment, devoted to its modern professionalism yet deeply ingrained with the splendour and tradition of the past.

There was no possible excuse to abolish the Argylls. The claim that they were the junior Highland Regiment was invalid as they had their origins as the Earl of Argyll's Regiment, the first regular Highland Regiment raised by the British Crown in 1689. In 1968 they were the best recruited, the most operationally experienced and the best known battalion in the British Army. Once again the best interests of the nation were ignored in the interests of bureaucratic expediency—or was it that there was something more sinister and more subversive behind this whole event? It was my firm belief that if Britain was to survive as a democracy into the Twenty-first Century she must restore her national spirit by stopping this constant erosion of the sound and solid virtues without which any nation must surely perish.

I retired from the British Army on 30th September 1968. A week later I was on my way to Vietnam as a civilian war correspondent for the London *Daily Express*. As my aircraft touched down at Saigon the United States Air Force were taking off on a combat mission, a rocket attack had taken place just outside the city and there was a smell of war in the air. This, I thought to myself, is terrific. It was where I came in—full of optimism for the future with a new career to look forward to and a challenge all the way—but I would agree with Dr. Samuel Johnson that 'every man thinks meanly of himself for not having been a soldier' and it was an experience I would not have missed—all twenty-five short years of it.

INDEX

243

244

245

Mitchell, Angus (son), 132
Mitchell, Colin (father), 17, 18, 19–22, 23–5, 26, 27–8, 60
Mitchell, Mrs Janet (mother), 17, 18, 19, 20, 21, 27, 28
Mitchell, Lieut-Colonel Colin: family background and schooldays, 17–26; early interest in Argylls, 17, 18, 26; scouting, 21–2; during Second World War, 23–49; joins Home Guard, 24–5; enlists, 26; training with Royal West Kents, 26–8; in 164 Infantry OCTU, 28; awarded Belt of Honour, 28–9; platoon commander training, 29; with 8th Argylls in Italy, 29–49, 52; first leads troops into action, 37–45; in Palestine, 50–66; first taste of terrorism, 50–2; permanent commission, 57; first commands own company, 63; lessons of leadership, 64, 66; leg wounds, 65–6, 67; as ADC, 67–9; with 1st Battalion in Korea, 69–89; temporary major, 81; Adjutant to 8th, 90; in War Office, 90–1; Staff College, 90, 91; marriage, 91; return to 1st as company commander, 91; in Cyprus, 91–107; failure of Operation 'Kingfisher', 96–100; in Germany, 108; with KAR in East Africa, 108–18, 122; with 1st Argylls in Borneo, 120, 122–31; rejects Joint Services Staff College course, 122–3; with Chief of Defence Staff, 132–41; views on defence, 134–45, 148–50, 154; visits Aden, 147, 150–1, 160–1; dismayed at attitude to terrorism, 150–3, 160–1; with 1st Argylls in Aden, 157, 167 *et seq.*; intensive training programme, 159–60, 161–167; philosophy of command, 159–160; policy directives, 159–60, 163, 217; 'Seven Principles' for internal security, 163, 217, 235; on 'brutality', 162–3; and massacre of 20 June, 1–16, 167, 169–70; and recapture of Crater, 169–87; display of confidence, 173; difficulties with superior officers, 173–4, 178, 184, 202–5; patrolling programme, 174–7; daylight reconnaissance, 175–7; advance into Crater, 179–87; reoccupation, 188–232; methods of controlling terrorism, 189 *et seq.*; showmanship, 193; suppression of nightly activities, 194: interrogations,196; 'smear' campaign against, 197–9, 208; told to 'throttle back', 198, 199; Battalion Order of 18

July, 199–200, 202–3; warned about, 203; encourages publicity, 208, 233–4, 236; HQ attacked by terrorists, 211–14; firm action against Armed Police, 229–31; evacuation of Crater, 232–5; criticisms of, 235–6; Honours omission, 239; removed from promotion list, 240; resignation, 240–1, 242
Mitchell, Colina (daughter), 132
Mitchell, Hetty (Penny), (sister), 17, 18, 20, 23, 25, 27
Mitchell, Lance-Corporal Hughie, 175, 223
Mitchell, Lorne (son), 132
Mitchell, Sue (wife), 3, 91, 96, 132
Moncur, Major John, 5, 7, 10, 12, 176, 186, 229, 230
Monro, Captain Hamish, 215
Monroe, Elizabeth, 154
Moores, Private Johnny, 5, 12
Mountbatten of Burma, Earl, 133–4, 136, 137, 147–8
Muir, Major Kenneth, 70, 71–2, 88
Mukeiras, 170
Munnion, Christopher, 15
Munro, Lieutenant Gordon, 42
Murray, Sergeant-Major, 86

Nairobi, 108
Naktong River, 70
Nanyuki, 111, 118
Nathanya, 55
Neilson, Lieut-Colonel Leslie, 69, 75
New Zealand Division, 2nd, 36–7, 40, 48
Nicosia, 92–3
NLF (National Liberation Front), 10, 149, 170, 171, 209, 215, 218–20, 221–3, 224–7, 230–1
Norbury, 17, 18, 20, 21, 28, 68

Oakley, Piper, 212
Odinga, Oginga, 109
O'Donovan, Lieutenant Brendan, 213
O'Dowd, Major Benny, 88
Orr, Lance-Corporal R., 201, 202, 206
Owen, Peter, 185, 186

Pa Lungan, 128, 130
Pa Umer, 129
Pakchon, 78, 79–81
Palestine, 50–66, 101, 152, 202, 212
Palmer, Major Patrick, 178, 180, 223
Paphos, 93, 96, 100–1, 161
Parachute Regiment, 10, 56, 144, 187, 233
Penman, Major John, 50, 63, 73, 76, 78, 79, 81
Perth, 29, 91